D1395505

CULTURAL BABBAGE

Cultural Babbage
Technology, Time and Invention

edited by
FRANCIS SPUFFORD *and* JENNY UGLOW

faber and faber
LONDON · BOSTON

First published in 1996
by Faber and Faber Limited
3 Queen Square London WC1N 3AU

Photoset by Datix International Limited, Bungay, Suffolk
Printed in England by Clays Ltd, St Ives plc
Set in 10/13pt Monophoto Photina

A CIP record for this book
is available from the British Library

ISBN 0-571-17242-3

10 9 8 7 6 5 4 3 2 1

To Julian Loose

Contents

Sir: In your otherwise beautiful poem 'The Vision of Sin' there is a verse which reads – 'Every moment dies a man, Every moment one is born.' It must be manifest that if this were true, the population of the world would be at a standstill . . . I would suggest that in the next edition of your poem you have it read – 'Every moment dies a man, Every moment $1\frac{1}{16}$ is born.' . . . The actual figure is so long I cannot get it onto a line, but I believe the figure $1\frac{1}{16}$ will be sufficiently accurate for poetry. I am, Sir, yours, etc., Charles Babbage

Babbage to Tennyson, 1851

Introduction: 'Possibility'

JENNY UGLOW

The whole history of this invention has been a struggle against time.
Charles Babbage, 1837

But, ah, Shakespeare! . . . My, but that man did have ideas! He would
have been an inventor, a wonderful inventor, if he had turned his
mind to it.
Thomas Edison, 1911

This is an informal book about form: about invention and con-
straint, scientific metaphors in poetry and politics, images and spectacle in science. It is concerned with possibilities, with machine
dreams and technology terrors. And because theoretical leaps and
ingenious inventions take place in time, it is also concerned with
process: what scientists might call sequences of experiment and
measurement, cause and effect – and writers might call stories.

The idea for the collection sprang from three sources, a machine,
a novel and an image. The machine is Charles Babbage's 'Difference
Engine', designed in the 1820s, built in various bits and versions in
the 1830s and 1840s, but only finally constructed in a plausible
shape in the 1990s. The novel is *The Difference Engine* (1988), by
William Gibson and Bruce Sterling, which plays with an alternative
history for the nineteenth century if Babbage's machines had been
developed. The metaphor is Ada Lovelace's complex – but clear –
description of Babbage's next great invention: 'The Analytical
Engine weaves algebraic patterns just as the Jacquard loom weaves
flowers and leaves.' Her image suggested a powerful merging of
ideas, a transgressing of borders between engineering and imagination, number, gender, nature and art.

If Sterling and Gibson's 'what if' re-creation of upside-down
Victorian England induces historical vertigo, so does the Science

Museum's engine: a Victorian machine which no Victorian ever saw. It can make a spectator of the 1990s feel like a tourist at the Great Exhibition. The Difference Engine was – is – extraordinary, as a mode of calculation, as a piece of precision engineering and as an aesthetic object. The numerical processing may numb the brain, but the delicacy of the gleaming double helices of steel, spiralling upwards as the gears crank the interlocking cylinders from units to tens, is startling and beautiful. Suddenly this architectural ton of metal slips from the realm of engineering and calculation into the numbered rhythms of dance, the abstraction of mobile sculpture. And if it breaks perceptual classifications, it also seems to breach the barrier that time erects between minds of different eras, an experience which its builders felt intensely, as Doron Swade, Curator of Computing, explained in a letter:

It *was* eerie seeing the machine materialize after having lived in the mind so long ago. Seeing the engine operate made it easier for us to understand old Babbage's resentments: he could visualize the elegance and beauty of its function but his appreciation was private. It's as though his personal Pygmalion was invisible to everyone else.

The studies that follow reflect different aspects of this slippage. They are 'illegitimate essays', in Gillian Beer's neat phrase – exploratory forays which let writers trespass outside their 'proper' subjects. As witnessed by the teasing response to Tennyson which forms the epigraph to this book, Babbage himself did feel an urge to reduce the world to number, but he was also a philosopher and a humanist. These essays, in different ways, follow his lead, combining research with speculation, analysis with play. Ballasted by detail but untethered by the ropes of academic apparatus, they let the balloon of ideas float free across the orthodox science/arts divide.

Writing on technology, we are already dealing with slippery terms, shadowed with a penumbra of associations, easily diverted from the practical to the social, or from the maker to the made. Invention, which has its semantic roots in finding 'a way in', the capacity for discovery, quickly came to describe an active mental quality, the imaginative framing of a plan, a work, an object. But it could also mean the thing made – the poem, the toothbrush, the typewriter. Invention is the physical incarnation of vision by means of technical skill, a requisite of art as much as of science, as Dryden notes in his dedication to his *Aeneis*: 'A poet is a maker, as the word signifies: and he who cannot make, that is, invent, hath his name for nothing.' To an eighteenth-century engraver like William Hogarth the greatest fear was being a mere copyist, part of a mechanical process, unable to express any individual vision. Hence his insistence on signing his prints '*W. Hogarth invenit*' or '*W. Hogarth fecit*'. To invent, to make – the verbs of technology – were, for him, the signature of the artist.

Edison could thus admire Shakespeare as a 'great inventor', a genius with ideas; and, conversely, could see himself as a poet of the practical. Discussing the relation between technology and imagination in America (D.Phil. thesis, York University, 1995), Portia Dadley draws attention to the way that language authenticates the work of the inventor: a design cannot be patented until a description of it is provided in writing. And when Jacob Bigelow wrote his *Elements of Technology* (1829), he revived an old seventeenth-century word to cover the great variety of 'practical applications of science':

I have adopted the general name of Technology, a word sufficiently expressive, which is found in some of the older dictionaries, and is beginning to be revived in the literature of practical men of the present day. Under this title is attempted to include an account ... of the principles, processes and nomenclatures of the more conspicuous arts, particularly those which may be considered useful, by promoting the benefit of society, together with the emolument of those who pursue them.

As Clive Bush pointed out in The *Dream of Reason: American Consciousness and Cultural Achievement* (1977), the root of technology – *techne* – means grammar: the articulation of certain principles. The 'syntax' of technology can be conceptual as well as physical, beautifully envisioned, for example, by Robert Fulton, the pioneering steamboat inventor: 'The mechanic should sit down among levers, screws, wedges, wheels etc. like a poet among the letters of the alphabet, considering them as the exhibition of his thoughts, in which a new arrangement transmits a new idea to the world.'

Technology transmits the 'idea' of invention: converts shadow to fact. In this connection, it's worth noting that Prometheus' real crime, and real legacy, according to Aeschylus, was not the theft of fire but the invention for humans of 'the art of numbering, the basis of all sciences, and the art of combining letters, memory of all things, mother of the muses and source of all the arts'.

But before it is made, an invention is a 'non-fact', a construction of letters and numbers only, fantasy not reality: when we call an explanation a 'mere invention', we mean a lie. That friction between positive and negative readings crackles through public attitudes to science, revealed as soon as we transfer vocabulary from the technical to the human. Words such as 'experiment' often suggest irresponsibility and risk rather than steady scientific testing. 'Mechanical' can be the common equivalent of 'soulless', while 'machine' slithers into the over-clever (and therefore evil) 'machination', or into 'the machinery of justice', 'the party machine', 'the Nazi war machine' – the dehumanizing of social institutions. Any examination of the culture of science takes place on an image-ridden battleground.

This book does not offer a continuous history of the battle or a comprehensive map of the terrain. It focuses instead on underlying attitudes and particular advances (on a lesser scale than the 'great steps for mankind' of agriculture, medicine or space conquest) which seem, in retrospect, to illuminate the cast of mind of their day, and ours. In doing so, it offers tentative explorations of the process by which scientific discoveries become inscribed on the collective and individual psyche, a route taken most notably by Isobel Armstrong in her reading of the cultural resonances of glass technology in the 1850s.

Far from aiming at a smooth, homogeneous whole, the essays are as individual in style as in content. But they spin a web of connecting themes: the point in research or design when the gulf between 'scientist', 'artist', 'philosopher' vanishes; the way the language of technology and avant-garde science provides metaphors for social organization and, conversely, the 'impurity' of pure science, which conceives and conveys its findings not only within the political and technical constraints but also within the linguistic and cultural terms of the era. Beneath these interrogations lie Babbage-like problems of scale, measurement, precision, the world as 'calculable'; the relation between chance and certainty, gambling and probability. These questions in turn are linked to the framing of philosophical perceptions of the universe and psychological notions of the self.

Because our starting point is machines – artifice rather than nature – the shadowy dot-to-dot pattern that emerges takes a slightly different shape from the more familiar 'evolutionary' picture, in which images drawn from the physical sciences are applied to historical and social process. In that 'natural' scenario, Chambers and Darwin take over 'God's plots', geology smashes theology and theories of evolution mutate into social Darwinism: 'survival of the fittest', 'natural selection' and territorial greed dictate the struggles of capitalism, commerce and empire. But evolutionary language has always run alongside the equally laden diction of design (Darwin's original projects would have been unthinkable without William Paley's theological arguments for design and adaptation in natural phenomena). And once writers draw metaphors from the applied, rather than the natural, sciences – invention, engineering, industry and power (in all senses) – then the organic model of an evolving society gives way to the constructed. Psychological, political and theological theories are framed in terms of control and social engineering ('they have designs on us'), and the images are those of machines, cogs, computers, rather than biological processes.

In the 1790s the language of experimental science fused easily with both Enlightenment thinking and utopian politics: the natural philosophers of the Lunar Society in the 1780s held to the Lockean faith that 'a man, by the right use of his natural abilities,' could attain a knowledge of God (a claim revived, in a rather different, neo-mystical form, by modern physicists like Paul Davies). By reason, Joseph Priestley maintained, one could find out God's laws and advance His plan. But Priestley's belief in necessity, that all things proceeded by means of immutable laws, both moral and physical, did not deny human responsibility and connection: the scientists' cry on the eve of the French Revolution, was 'Liberty, Reason and Brotherly Love for ever!'

To opponents such as Edmund Burke, Priestley's reasoning, like the gases he experimented with, was inflammable stuff, at once danger-ously uncontrollable and also too determined, too measured. Tom Paulin, Neil Belton and Jon Katz all see Burke as the great progenitor of the British scorn of science, initiating a cultural (and political) divide which persists in the identification of 'poetry' with conserva-tive romanticism and sensuous feeling and of 'science' with ruthless rationalism. Behind this division lurks Hazlitt's divisive definition in his essay on *Coriolanus*:

The imagination is an exaggerating and exclusive faculty: it takes from one thing to add to another: it accumulates circumstances to give the greatest possible effect to a favourite object. The understanding is a measuring and dividing faculty; it judges of things not according to their immediate impres-sion on the mind, but according to their relations to one another . . . The one is an aristocratical, the other a republican faculty.

One might add, though – *pace* Paulin and Belton – that the republic of science, in Hazlitt's day, would have been a republic ruled by ironmasters and merchants, not a proletarian idyll. Weavers were described as 'hands' a good decade before Adam Smith's *Wealth of Nations* gave a mathematical precision to the 'laws' of production and the self-interest of the market. Priestley was a realist as well as an idealist, accepting that it was capital, as much as conviction, that fuelled scientific experiment. It was not only experimental science that called down the wrath of Burke and his Romantic-conservative heirs, but also the appeal to theoretical models, easily transferred to institutions which they felt were by nature 'organic'. Number-crunching proto-Babbagery, in fact.

Despite the Burkean backlash, the inventive fever and numerical passion of the late eighteenth century continued: the discoveries of Humphry Davy flowered in the era of his assistant, Faraday. In *On the Definitions of Life Hitherto received: Hints towards a more Comprehensive Theory*, published in 1848, Coleridge looked back excitedly, first at the progress from Kepler and Newton to Hartley, when 'not only all things in external nature, but the subtlest mysteries of life and organization, and even of the moral being, were conjured within the magic circle of mathematical formula', then to the discovery of electricity, which had 'electrified the whole frame of natural philosophy'; and finally to the systematization of chemistry by Lavoisier. The mood was euphoric:

Henceforward the new path, thus brilliantly opened, became the common road to all departments of knowledge: and, to this moment, it has been pursued with an eagerness and almost epidemic enthusiasm which, scarcely less than its political revolutions, characterize the spirit of the age.

This was an age whose inventors' names, long divorced from their possessors, endure in our electrical labels, our laboratories and schoolrooms: Galvani and Watt, Volta, Ohm, Ampère, and later Bunsen, Joule and Kelvin. But in the decades between 1800 and 1840 'natural philosophy' was still open to all. (William Whewell's 1833 coinage of the word 'scientist' began to signal its inevitable professionalization, but the term did not get into popular currency anywhere until the 1860s – Faraday always refused the label.) Within the Nonconformist communities, shaped largely by Priestley's vision, scientific inquiry took institutional shape in the Literary and Philosophical Societies of the 1790s and the Mechanics Institutes of the 1820s. 'Gentlemanly' experiment continued, and the professors laboured on in leisured isolation, but the most significant aspect was the democratic ethos of experimental work. Women attended popular scientific lectures in the same numbers as men. In practical research, class was not an issue: experience was what counted, quick thinking on the ground, the tactile knowledge of men with oil on their hands rather than kid gloves, the tradition of Watt, Stephenson and Trevithick, or of William Sturgeon, bootmaker and soldier turned lecturer, who invented the electromagnet in the 1820s.

This mode of invention was a function of the workplace, of commerce. Technology has never quite lost its boiler-suit status, in comparison with white-coated theory and 'pure research'. (Interestingly, as Marek Kohn reported in *The Independent*, some scientists today hold that the West is too theoretical, that we need to return to the experimental, pointing out, for instance, that James Watt's steam-engine governor worked for many years before James Clerk Maxwell proved its stability mathematically. The fuzzy logic currently being applied to domestic technology and computing takes its model from this workplace style, reflecting the human habit of graded adjust-ment – 'just round the corner', 'a little to the south', 'try pumping it harder'. When fuzzy maths are used in conjunction with the new neural computing – machines that can learn – we may have some-thing resembling a clever apprentice.)

In Britain, as the notion of a professional, academic élite took hold, this democratic strand in scientific work became a problem, an issue reflected in the memorable words of Sedgwick, President of the Association for the Advancement of Science, in 1842. Walking through the back streets of Manchester, he was astounded, he said, that, 'in talking to men whose brows were smeared with dirt and whose hands were black with soot, I found in them the marks of intellectual minds and the proof of high character.' But while he urged his fellows to look beyond entrenched prejudice, he also added hastily: 'Do not suppose for a moment that I have any levelling doctrines.'

With each new wave of technological advance, the rhetoric of social possibility, and the reaction to it, resurfaces. Sometimes the lan-guage is moral, sometimes political, sometimes reformist, sometimes genuinely democratic. Marina Benjamin's account of microphotogra-phy makes us realize that as the 'parlour microscope' of the 1850s meant everyone could be a scientist with access to the 'secrets' of life, so the notion of extending the public's 'range of vision' acquired aesthetic, theological and political overtones. Similarly, Isobel Arm-strong points out that glass arcades were features of social as well as physical architecture: Fourier envisaged them as 'one of a thou-sand amenities kept in reserve for the new social order'.

One recognizes the same visionary drive today: the parallel to the 'parlour microscope' is the 'home computer', which promises a simi-lar entry to knowledge, a similar sense of connection, and has the same centrifugal force. We love technology that offers to bring the whole world to *us*, creating Faustian illusions of centrality, participation, control: whether it be the microphotographs of Victorian vistas, the imperial inclusiveness of the Crystal Palace or the World Wide Web which brings the Louvre onto our screen. The cancelling of distance is part of the appeal, the access to the exotic and 'hidden'. Writing in the *Guardian*, Jack Schofield lamented that while the Net told us the forecast for Hawaii, or the last bus in Boston, it might be 'a damn sight more useful' if we could find out the weather in Manchester or the last train from Euston. Well, in fact we can, but it's a damn sight less of a buzz.

The buzz that surrounds the Net recalls Coleridge's heady words of the 1840s, linking scientific and political revolutions. To begin with, the Net seemed like power to the people at last, an eager civilian takeover of military, state and corporate encampments. The organizational principles of the World Wide Web, with its loops and internal linkages, cut through hierarchies of knowledge, as well as control. You could enter at your own level and discover the world. Hazlitt's aristocratic and republican imaginations fused at last: in terms of learning, the journey, not the arrival, really did matter. One could even see the Net as a space, rather than a system, a physical realm of encounters, communities, buildings, explorations.

Self-assertion, even the imprint of domesticity, could have its place in the 'Home Pages' with their hot-spot coffee cups and teddy bears (or, if you're Clinton, a miaowing image of Socks, the White House cat). But the new sense of connection – between users, between hitherto separated areas of knowledge, between scattered parts of the globe – created the thrill of a collective enterprise. This was all for one and one for all, and all for the price of a modem and a monthly sub. A digital utopia was at hand: politics would be transformed since all citizens (or all those on-line) could join in the debate. Errors of planning would no longer happen, since we could work out the consequences in advance.

This strand is exuberantly evoked by Jon Katz's essay on Tom Paine. The Net offers a natural home to electronic anarcho-syndicalists, or to optimists like Sadie Plant, who aligns the liberation of

'thinking machines' with a new world order and particularly with
the empowerment of women. Women, Plant argues, have so far
been machines for men: now the computer, whose operational mode
of interconnection, knitting, patchworking and chat 'naturally' fits
women's minds and life patterns, could put a time bomb under
patriarchy.

So fast are things moving that these visions of harmony, sharing
and grass-roots subversion are already under threat. Traffic lights
are going up on the digital superhighway. Patent cases are affront-
ing the Net community: 'How can they lay down laws to say what
I can have on my own hard disk?' Beneath the exhilarating whoops
of egalitarianism rumble the (equally levelling) rumours of international viruses, global crashes, pirates and wreckers, mind benders
and markets. Some analysts argue that the interactive intimacy of
cybernetic human/machine involvement is in itself desensitizing,
that it limits rather than extends awareness. During the Gulf War,
the resemblance between computerized film of missile attacks and
games led TV viewers to complain of the grainy quality and poor
graphics. More significantly, the habit of cybernetic game involvement, the urge to cry 'Score! Hit!', could overrule the horror of the
real: the only thing missing was the remote in our hands.

Game patterning drives many of the new extensions of this technology, a daunting thought in view of the expanding military use
of computers for strategy and tactics, as well as weapons research.
And the language of the Net – like its name – is also the language
of dependency, surveillance and control. The utopian visions that
cluster round new technologies are invariably shadowed in this way
by dystopian counterparts. Science has long been dogged by theological or mythic alarms – the hubris of interference with nature, intervention in 'God's work'. (Even Babbage had wizard-like flights of
fancy, spending a long time perfecting shoes to walk on water.) Our
perception of scientific knowledge is unstable. It fractures and splits
between notions of reason and magic, the sane and the arcane, the
life-enhancing and the mass-destructive, the 'modern' and the
Gothic. Thus in nineteenth-century Britain the old myth of Prometheus became inextricably linked in the public mind (as in her title)

with the new myths of Mary Shelley's *Frankenstein: Or, The Modern Prometheus*: the creation of an 'artificial man' and the 'mad scientist'. And the monster, the inhuman human, also became a synonym for the ungoverned, angry, parentless populace to which science and industrialism seemed to have given birth.

By contrast, in America the power of scientific optimism was stronger, boosted by a more positive (if equally fallible) set of myths: the rational, egalitarian constitution; the ever-retreating frontier. Instead of Frankenstein's monster we have 'the self-made man'. Technology could break old bounds, spiritual as well as physical. In the early 1840s Emerson could write that 'Machinery & Transcendentalism agree well. Stage Coach & Rail Road are bursting the old legislation like green withes.' In these wishful narratives, America itself is the appliance of science, a state built on first principles, the concretization of a dream. The inventor's workshop could be a technological Eden; his self-interest could be seen as altruism and his protective secrecy as sound commercial sense – the stuff that built America. The right stuff. But the States, too, had their blacker myths and Romantic reactions: the landscapes of Poe, the burnt-fingers pessimism of Twain, the fatal, overreaching experiments of Aylmer in Nathaniel Hawthorne's 'The Birth Mark', frustrated in his search for the laws of Mother Nature, who seems so generous and open yet guards her secrets so fiercely, 'like a jealous patentee'. In *Civilizing the Machine: Technology and Republican Values in America, 1776–1900* (1976), John Kasson points out that fifteen years after his euphoric journal entry Emerson was placing industry in a more sinister light, noting its power to enslave as well as liberate. Indeed, Emerson hints that the American love affair with invention springs from the same impulse as the use of slavery, a desire for reward without effort:

I cannot accept the railroad and the telegraph in exchange for reason and charity. It is not skill in iron locomotives that makes so fine civility, as the jealousy of liberty. I cannot think the most judicious tubing a compensation for metaphysical debility.

This fierce address at Concord in 1851 has a bitter, premonitory tone. Slavery would extort its price: 'the greatest prosperity will in vain resist the greatest calamity'. A fundamental tension remained between the democratic dream of free invention as a social good – liberty, reason, brotherly and sisterly love – and the nightmare of manipulation, possession and exploitation, of hubris and disaster.

Machine fears take theological, philosophical, social and psychological forms, and inevitably these overlap. Thus the idea that the theological, determined, 'universe machine' denies free will becomes a commonplace, an easy joke, as in the old limerick:

> There once was a man who said, 'Damn!
> It is borne upon me that I am
> An engine that moves
> In predestinate grooves
> I'm not even a bus, I'm a tram.

Translated into the relationship between the individual and society, this fear of being controlled or of being denied liberty has always coexisted uneasily with the belief that science *gives* liberty. Even the pursuit of knowledge, the priority given to accumulation of facts, could be seen as life-denying. Like Shelley, Coleridge and Hazlitt, Wordsworth was excited by Priestleyan experiment. 'The Man of Science' as spiritual pioneer is greeted with zeal in the Preface to the *Lyrical Ballads* in 1798. But within a few years, in the 1805 version of *The Prelude*, he attacks the kind of intellect whose 'globe and sceptre' are 'telescopes and crucibles and maps' (as opposed to 'Nature's book' or the imaginative flights of the *Arabian Nights*) and warns against:

> These mighty workmen of our later age
> Who, with a broad highway, have over-bridged
> The froward chaos of futurity
> Tamed to their bidding . . .
> . . . the tutors of our youth,
> The Guides, the Wardens of our faculties
> And Stewards of our labour, watchful men

And skilful in the usury of time,
Sages who in their prescience would control
All accidents, and to the very road
Which they have fashion'd would confine us down
Like engines.

Wordsworth's swerve from progressive to conservative is typical of a deepseated nervousness about the relationship between science, number and technology and the 'freedom of the individual mind'. On the one hand, in the eighteenth century it had seemed blithely liberating to cast aside superstition and to see man and the state as machines, capable of regulation, amenable to tinkering. On the other hand, if men were machines, created by 'the system', then the 'masters' could also dismiss them as broken down, with a screw or two loose. Thus in William Godwin's novel *Caleb Williams* (1794), Mr Collins says loftily to Caleb:

'I regard you as vicious; but I do not consider the vicious as proper objects of indignation and scorn. I consider you as a machine: you are not greatly constituted, I am afraid, to be greatly useful to your fellow men: but you did not make yourself; you are just what circumstance irresistibly compelled you to be.'

When the model of specific, limited operations in which each individual act was defined by an overall 'system' was enshrined in industrial organization, many commentators glowed with enthusiasm for the efficiency of the 'factory machine'. But at the same time others, like Carlyle in *Past and Present*, fulminated against its alienating force. In political terms, the mechanical slur has often been levelled at totalitarian, centralized states. But it could seem different from the socialist perspective. 'Under capitalism,' Lenin wrote in *The State and Revolution*, 'we have a state in the proper sense of the word, that is, a special machine for the suppression of one class by another.' William Morris, writing to the editor of *The Standard* in 1883, used an image which identified liberty with technological *un*-making, affirming that, 'we are but minute links in the immense chain of the terrible organization of competitive commerce, and that only the complete unrivetting of that chain will really free us'.

The onslaught against society as mechanism intensified in the West in the 1920s in response to Taylorism, corporatism and notions of 'scientific management'. Yet the fear was complicated by the fact that societies *wanted* the very machines whose operation might condemn them to an automaton-like existence. This sense of fearful seduction, as Ludmilla Jordanova points out in *Sexual Visions* (1989), is implicit in the way the inventor in Fritz Lang's *Metropolis* (1924) makes his robot a woman 'who lures men through desire' – called 'Maria', mother of God. The real Maria embodies the heart which the city, and its enslaved male workforce, need; her robot image – sexuality without love – threatens to destroy them. (Any suspicion of male-imposed fantasy is muddied by the fact that the novel on which the film was based – like *Frankenstein* – was written by a woman, Lang's wife.) This strand endures in current sci-fi fantasies, in which terror of artificial beings, whether holograms or cyborgs, is induced not by their inhuman ugliness (as with Frankenstein's monster), but by their beauty and strength, their muddling *likeness* to a human ideal. The problem becomes one of authentication, of defining the human – or, conversely, of falling in love with the feared machine.

The 'Metropolis' spectre of industrial society as a machine which crushes or enslaves us also continues to the present. The cyber counterpart of 'a mere cog' could be being a mere chip, but an added fear is that of being written into the software as a digit, a number, a unit of information – not its controller, or even its processor. By contrast, one can detect a countermood, not utopian but nostalgic, affecting the current dynamic of old versus new, tradition versus modernism. This reading actually mourns the gritty industrial technology of the past, using the machine as metaphor for lost communities and disregarded skills. In 1993, Jeff Torrington's *Swing Hammer Swing* offered a virtual lament for the Gorbals, its demise as a living community symbolized by the gravestone-like slabs of 1960s high-rise blocks:

What set the red nerves twitching was the utter contempt for the working classes which was evident no matter where the glance fell. Having so cursorily dismantled the community's heart, that sooty reciprocating engine, admittedly an antique, clapped-out affair, but one that'd been nevertheless capable of generating amazing funds of human warmth, they'd bundled it off into the asylum of history with all the furtive shame of hypocrites dumping Granny in Crackpot Castle.

All these examples, cosmic or social, impinge on individual psychology. Macro defines micro. Those of us who live in 'developed' nations don't quite like being made to face the fact that our lives are governed by technology from the antenatal clinic to the crematorium, or that our cities are built on a Gruyère cheese, its holes threaded by electric cables, sewers, gas and water pipes, while our skies are a moving web of communications, planes and satellites.

Alongside the feeling of power and expansion that science can bring goes the inescapable Pascalian tremor which comes from the recognition of human smallness. If knowledge is power, it can also induce impotence. The telescope that takes us to the ends of the universe presents Earth as a dot; the microscope that reveals a single cell crumbles 'independent' life-forms into bundles of competing organisms; the current possibility of storing information on a single organic molecule instead of electronic circuitry will condense vast libraries of memory to an element a billionth of a centimetre across. And, as Alison Winter and Anne Joseph argue, the process which displays our essential identity – like genetic fingerprinting – also reduces that identity to an invisible strand of protein. The physicist peels off the layers of matter – molecular, atomic, nuclear, hadronic and quark – until all we are left with is pure energy. Where is the self?

'Who are *you?*' said the Caterpillar.

This was not an encouraging opening for a conversation. Alice replied, rather shyly, 'I – I hardly know, Sir, just at present – at least I know who I *was* when I got up this morning, but I think I must have been changed several times since then … But if I'm not the same, the next question is "Who in the world am I?" Ah, *that's* the great puzzle!'

Alice's identity crisis can be phrased in different ways, but it inevitably throws up the issue of mind–body dualism. In this debate
consciousness is sometimes seen merely as a matter of physical
stimuli, an act of sensation rather than an expression of 'soul', and
the idea of the body/mind entity as an organic mechanism may
lead on to the view of an individual *as* a machine (which is subtly
different from seeing people as *part* of a machine). Such discussions
hark back to La Mettrie's *L'Homme Machine* (1747), which attempted (partly provoked by Descartes' definition of animals as
'automata') to restate the problem of mind as physical, not metaphysical – to define man as a mechanical entity in which
perception, emotion, understanding, foresight and thought were produced by organic causes. The early years of cybernetics, not surprisingly, were haunted by the ghost of La Mettrie, as works such as W. R.
Ashby's *Design for a Brain* or Norman Wiener's *Cybernetics; or Control and Communication in the Animal and the Machine* suggested that
self-directive force does not contradict but may even be derived
from mechanical causation. In 'Babbage's Dancer', Simon Schaffer
describes how automata, dancing dolls and chess-playing Turks exercised a fascination for Babbage and his peers which went beyond
the admiration of intricate mechanism to touch the question of
machine intelligence, and therefore human intelligence. And he
suggests, too, their deep need to locate the manipulators of 'superhuman' intelligence elsewhere, as a threatening Other, alien to
their society, associated with the inscrutable Orient.

This type of displaced fear is recognizable today in the West's
ambivalence towards the Pacific Rim. In the late 1980s the realization that the US nuclear arms industry was now dependent on
Japanese parts induced a wave of paranoia. If technology was
synonymous with 'the key to the future' then, by definition,
America no longer led the world. 'Sake imperialism' could replace
'Coca-Cola colonization', as David Morley and Kevin Robins put it
in *New Formations* (1993). One reaction was to recast fears of
the machine world in racist visions of the Japanese as 'aliens' or
replicants, and to see Japanese society as oppressively automated,
suppressing the 'soul' (and thus likely to be fissured by the hysteria

of religious cults, a Romantic reaction from the depths). A different response – that of Ridley Scott's *Blade Runner* or William Gibson's *Neuromancer* trilogy – was to make Japan the site for the most intense exotic and erotic fantasies of postmodern society, a Nintendo world in which the real and the electronic blur, frightening and alluring at once. Both approaches, however, insist on the 'inhuman' element in technological dominance.

As illustrated by this West–East alarm, there is a danger of seeing the history of technology – a narrative of endeavour and excitement – as a tale of global conspiracy, a constant snatching away of power by an élite group or nation. Spears and hoes are followed by arte-facts which endow 'magic powers' and authority; literacy is followed by bureaucracy; metallurgy by industry and conquest. True, as James Burke and Robert Ornstein point out in *The Axemaker's Gift* (1995), if science has changed the world it has not always been for the better, as the inequalities of the planet show. (State intervention is rarely altruistic: the UK government's promises of R & D invest-ment, following the 'Technology Foresight' report on key technolo-gies for the twenty-first century, don't highlight the human benefits but the need to 'keep Britain at the forefront of world science, engi-neering and technology'.) But although we may respond to science with ambivalence, dubious of its power, its ethics, its nationalism, its version of progress, its traditional mechanistic slant, this should prompt us to explore the implications and seek democratic means of control, not to turn our backs.

On a lighter note, it isn't always possible to take national plan-ning, or the machine as a model of social or economic organization, altogether seriously. In theory, such models should 'work', and pre-dictive machines have actually been built, on Stalinist lines in the Soviet Union and Keynesian principles in the West. The Science Museum recently displayed a restored Phillips Economics Compu-ter, which wowed the professors when first demonstrated at the London School of Economics in 1949. As Doron Swade describes it, 'The pumps were switched on and coloured water sloshed through tanks, pipes, sluices and valves. The levels settled, pulleys turned and a pen plotted and traced results. The machine was a hydraulic model of income flow in the national economy.' In 1949, flurried LSE students, attempting to control trickles of income and cascades of expenditure, played the roles of Chancellor of the Exchequer and Governor of the Bank of England, a touch of realism being added by the instruction to ignore each other's moves. The results were – predictably – disastrous. (People who shun mechanistic models can also take comfort in the fact that the Phillips Economics Computer proved temperamental and prone to leaks.)

Historically we have used chance for prediction – throwing dice or knucklebones, casting lots. One of the great psychic changes between the eighteenth and nineteenth centuries was the switch from a gambling-crazed society to an actuarial one, a Gradgrind world, its forecasting ruled by statistics. (Although Ada Lovelace was an addicted gambler, she did try to work out a system.) The twentieth century has swung away from notions of certainty, as physics revealed oscillation rather than stability and the precision of high mathematics cast up variation, imprecision, chaos. Chance rules again.

These kinds of conceptual shifts from stability to instability and back are illustrated by Gillian Beer's essay on wireless and modernism, which traces the way wave theory and quantum physics (broadcast across the 'airwaves') affected the politics, aesthetics and self-perception of artists, writers and 'ordinary people'. Complementing this approach, in 'Unstable Regions' Lavinia Greenlaw follows the territorial conflicts between science and poetry, their jealous guarding of differences, their borrowings and mutual misreadings.

Public attitudes are complicated not only by the various readings of the world in 'scientific' terms, whether biological or mechanical, but also by other factors, two of which might be defined as time and translation. In the form they are understood by the uninitiated, technological models, scientific discoveries and theoretical leaps are always, in some way, distorted. Physicists complain that the hardest part of their work is that of communication, of conveying mathematical abstraction in impure, emotionally nuanced language. But with increasing subdivision (a recent count throws up 90,000 specialisms), scientists have to 'translate' even to talk among themselves, adopting the techniques of poetry, storytelling and art. Burdened with reporting a symposium of high-level cosmologists in May 1995, Tim Ferris of *The New Yorker* returned constantly to the theoreticians' own visual and verbal formulations. Often the appeal (famously used in the case of DNA) is to aesthetics. Thus Edward Witten, defending string theory: 'When a theory is fantastically beautiful and physical – and string theory is – I don't find it plausible that it would be wrong.' Or Andrei Linde, illustrating the 'bubble universe' with colour graphics, including a spectacular 'Kandinsky universe'. Or Renata Kalosh, demonstrating the idea of 'superhair' in extreme black holes with sketches comparing a bald man to a punk with spiked hair-do.

Ironically, as the arts are called on to elucidate the sciences so our metaphorical understanding of science slips back from the precision of laws and mechanism towards mystery and magic: this book is scattered with examples of what Marina Benjamin calls 'the language of the laboratory leaning on the language of the fairground'. The model, the toy, the picture offer an intermediary stage, a mediating glass. This is an old tradition, to woo with spectacle – as Edison did, and the early electricians. Nineteenth-century scientists did not initiate the technology of display – think of the elaborate stage mechanisms of Baroque masques, or the demonstration devices in the George III collection at the Science Museum – it's rather that they turned the function of display around, making 'shows' into experiments. Reconstructing the war between the London Society of Electricians and the professoriate in the 1830s (in 'Currents from the Underworld', *Isis*, 1993) Iwan Rhys Morus quotes a splendid description of the shows that men like Sturgeon offered in the Adelaide Gallery:

Clever Professors were there, teaching elaborate science in lectures of twenty minutes each. Fearful engines revolved and hissed, and quivered. Mice led sub-aqueous lives in diving bells. Clockwork steamers ticked round a basin perpetually to prove the efficacy of invisible paddle wheels. There were artful snares laid for giving galvanic shocks to the unwary.

Spectacle (free from galvanic shock) still retains its appeal as a form of scientific communication – in the heightened pictures of space exploration or the images of cells. The baffling seduction of mechanical toys persists too, as in Kasparov's long-running chess battle with the Pentium PC and its Genius 2 software. Again and again we find the phenomenon implicit in Simon Schaffer's essay on automata, that technological inventions 'take off' in public imagination, once you can see their effects or use their power, but you don't have to understand their workings: their charm depends on the invisibility of the human skills that made them. At the turning point, we talk of 'thinking machines', 'cunning engines' and 'smart cards', but after a certain stage it is not the technology but the human use of it which is regarded as 'clever', a matter for praise.

Some inventions – like Edison's phonograph or Loudon's glass conservatories – are immediately appropriated in this way, burgeoning rapidly to symbolic status, sometimes (as Alex Pang reveals in his history of Buckminster Fuller's domes) in ways that reinvent even their inventors. Others seem at once in tune with and at odds with their age, like Concorde, a beautiful, expensive, politically importunate siren, dipping her beak in the fountain of time. In such cases the imaginative appeal sometimes seems to sail far beyond its practical or political context: as the critic Ian Hunt noted in a recent talk on anachronism, the political and cultural ironies multiply, for example, when capitalist high tech marries with tribal art:

Who is in control of the irony when a Concorde is remade in a South African township in sheet metal and rivets, like the one that hangs in the Horniman Museum in South London? The fact that the plane never flew there, that it was a mistaken idea of the future, does not obliterate the uncomfortable inequalities that lie behind this homage to a future that now will not be, from a continent from which capital flight is the salient political fact.

A different sense of anachronism surrounds inventions like the Difference Engine, which never come into full use in their own day but which contain ideas that lie like yellowing blueprints in dark cupboards, to be stumbled upon afresh by later generations. One thread in the web of this book is the fecundity of 'cul-de-sac' research or invention, the curious underground life of ideas which become a kind of 'science fiction' in themselves. And, like science fiction, they often 'come true': Clifford algebra, thought an eccentric dead end in the 1870s, is now used in the calculation of quarks; Thomas Bayes's theorem of 1763, assessing the probability of linked events, was employed by the artificial neural network that won 'The Great Energy Predictor Shootout' in Atlanta in 1992, to predict energy consumption in heating systems, paving the way to 'smart buildings'.

While it's impossible to divorce an idea from the cultural ethos which generates it, it can seem to exist outside time. In this respect, the fierce, embattled confidence of Babbage is curiously like that of his contemporary, Edgar Allan Poe. In 1848, in *Eureka*, Poe defiantly pronounced his theory of an evolutionary cosmos, a finite 'Universe of Stars' in which *'Space and Duration are one'* and matter is reduced to energy. So sure was he that he had cracked the code of the cosmos that he asked for a huge advance and a first print run of 50,000 copies. His publishers offered $14, printed 500 copies and sold fewer. Poe was thought mad, doped or drunk. A century later, just as computers vindicate Babbage's dreams, so relativity, big-bang theory and subatomic physics collude with Poe's wild surmise.

Although Poe's work looks prophetic, it was securely anchored in the gravitational laws of Newton and the cosmology of Laplace. Where he was truly revolutionary was, first, in defying the scientific notion of absolute 'axiomatic truth' apart from existing systems of thought, and secondly in placing empiricism below 'intuition'. To Poe, intuition was *'the conviction arising from these inductions or deductions of which the processes are so shadowy as to escape our consciousness, elude our reason, or defy our capacity of expression'* (his emphasis). Here, and in his fiction, he still demands rigorous logical reasoning, the 'Calculus of Probabilities', but, as Harold Beaver pointed out in 1976, in his fine annotations to the science fiction, Poe takes this to the point where mathematics and mysticism fuse. Fusion is an appropriate image.

Francis Spufford's essay delves into our preoccupation with these processes, the imaginative apprehension of science and its social consequences, the impulse of the great 'perhaps', the counterfactual doors to alternative pasts and potential futures. Historians make use of this approach – for example in estimating what the course of the Second World War would have been without the mathematical code breaking of Alan Turing and 'Ultra' – but fiction gives a chance to play and replay variations on given sequences, applying inventions which are only in the mind, resetting the equations of development and computing the outcome. Such fantasy, like Poe's, is a philosophical mode.

Often, in looking at scientific and technological development, the most interesting stage proves to be the time between 'invention', as idea, and 'an invention', the resulting product, the final equation. The social parallel is the stage when the consequences of a technological advance seem undefined and open, leaving space to envisage either positive or negative eventualities. In a recent fictional completion of Babbage's Analytical Engine, Peter Ackroyd's *Dan Leno and the Limehouse Golem* (1994), which connects the obsession with mechanical calculation directly to Benthamite utilitarianism, the engine attendant says, 'I know that with notation we might take away all sorrow.' Outside fiction, in both technical and societal spheres this gap of possibility is hedged in by political imperatives, financial constraints, the sheer difficulties of construction or the recalcitrance of logic. Reason gives way to luck, to circumstance, to will-power and, sometimes, to moments of dazzling insight. Running beneath the mundane slog of scientific research and development, beneath the long hours bent over test-tubes, fiddling with screws, evaluating printouts, glitters the same vein of imaginative creativity that we connect with artists and writers and call 'originality' (hard to be more 'original' than Darwin, Maxwell or Einstein). And outside the confines of the impersonal scientific paper, the reproducible experiment, the reasoned proof, it is interesting how often scientists talk about *un*reason and doubt. 'I know this business is free of contradictions, yet in my view it contains a certain unreasonableness' (Einstein). 'We take it for granted that it is perfectly consistent to be unsure' (Richard Feynman). 'Maybe God is fooling me ... but it makes life so exciting' (Andrei Linde).

It is the gap between hunch and final theory, dream and invention, that seems most full of possibility. In the words of Alan Kay, whose 1968 vision of the portable computer, an environment where children could learn by 'making', seemed, at the time, wholly unrealizable: 'You can go on working on an impossible project for a long time if it has a lot of romance to it.'

Serbonian Bog and Wild Gas:
A Note and a Pamphlet

TOM PAULIN

A gulf profound as that Serbonian bog
Betwixt Damietta and Mount Casius old,
Where armies whole have sunk: the parching air
Burns frore, and cold performs the effect of fire.
Thither by harpy-footed Furies haled,
At certain revolutions all the damned
Are brought: and feel by turns the bitter change
Of fierce extremes . . .

Milton, *Paradise Lost*, Book II

Science provided some of the most volatile material in the culture wars of the late eighteenth century. Edmund Burke hated science, identifying it with radicalism and godless republicanism. His *Reflections on the Revolution in France* (1790) is an attack on the modern, scientific principles which were a central part of the culture of rational dissent. William Hazlitt was formed by that culture (his father, the Revd William Hazlitt, was an Irish Unitarian clergyman who held advanced Whig views), and his argument with Burke was lifelong, passionate, full of a bitter, betrayed anger and an intense admiration.

From being one of the parliamentary champions of dissent, Burke became in his last tragic decade the enemy of Unitarian culture, attacking dissenting science, economics and the new subject, statistics, in particular, as enemies of that organic social hierarchy which he championed against those who sought to reform Britain. At one point Hazlitt speaks of the 'Serbonian bog', the 'rotten core' in Burke's understanding, making a witty, even racist hit which plays on the malicious perception of him as a sinister Irish Jacobite and secret papist who had been educated by the Jesuits.

In fingering Burke as an Irish Catholic, Hazlitt is doing more than minister to a particular prejudice – he is recognizing that *Reflections* is on one level a diatribe against the advanced Whiggish Protestantism Burke had espoused for most of his career. It is a diatribe occasioned by a sermon delivered by a leading English Presbyterian minister open to Unitarian thinking, Dr Richard Price, who was a friend of Benjamin Franklin, Joseph Priestley and of Hazlitt's father.

Price, who was a distinguished statistician and economist, delivered his 'Discourse on the Love of our Country' on 4 November 1789 at the Old Jewry, the Unitarian meeting house where Wordsworth would hear Joseph Fawcett (the Unitarian minister Hazlitt admired) preach the sermon which inspired part of 'Tintern Abbey'. In this sermon, which rejoiced in the French Revolution and asserted that the King of England owed his throne to the choice of the people, Price referred to an 'ardor for liberty' which had undermined superstition and error, and concluded:

Tremble all ye oppressors of the world! Take warning all ye supporters of slavish governments, and slavish hierarchies! Call no more (absurdly and wickedly) REFORMATION, innovation. You cannot now hold the world in darkness. Struggle no longer against increasing light and liberality. Restore to mankind their rights; and consent to the correction of abuses, before they and you are destroyed together.

Price's sermon was published by the Revolution Society which had been founded in 1788 to commemorate the centenary of the Glorious Revolution of 1688. The sermon angered Burke, who disliked the Society's anti-popery which he associated with Lord George Gordon's Protestant Association. (Gordon had told an angry crowd in June 1780 that Burke was behind the House of Commons' refusal to consider their petition for repealing the Catholic Acts.) As Conor Cruise O'Brien points out in his study of Burke, *The Great Melody* (1992), Price's pamphlet placed the British welcome for the French Revolution firmly in 'a context of anti-popery'. Price, like Priestley, was also a protégé of Lord Shelburne, who was Prime Minister from July 1782 to February 1783: Burke suspected Shelburne of having helped to bring about the Gordon Riots which followed from the rejection of the Protestant Association's petition.

Burke read Price's sermon in January 1790 and used it as evidence against any further progress in allowing religious toleration. In a speech delivered in the House of Commons on 2 March, he attacked Priestley, cited Price's sermon and raised the spectre of the Gordon Riots. Two years later, in a debate on Charles James Fox's motion for the 'Repeal of Certain Penal Statutes Respecting Religious Opinions', he attacked the Unitarians' petition as being against 'the general principles of the Christian religion, as connected with the state'. The Unitarians were the 'avowed enemies' of the Anglican Church and it was well known that Dr Priestley was 'their patriarch'. Burke then attacked Paine and linked the Revolution Society with the Jacobin Club in Paris.

Burke was therefore an avowed enemy of the Unitarian culture which shaped Hazlitt, whose relationship to his writings must be partly understood in this light. In *Reflections* Burke attacks Price and Priestley vehemently, mobilizing anti-Semitic prejudice against the Unitarians by ringing changes on the name 'Old Jewry', and in another nasty moment comparing Price to the Revd Hugh Peters, the parliamentary chaplain who was executed for treason at the Restoration.

The chairman of the Revolution Society was Earl Stanhope (1753–1816). Stanhope, a member of parliament, was a notable experimental scientist and radical politician who called himself Citizen Stanhope, sympathized with the French Revolutionaries and opposed the war with Revolutionary France. He experimented with electricity and invented two calculating machines, as well as a kind of printing press, a microscope lens, a stereotyping machine, a steam carriage and a special type of highly durable cement. Burke's hatred of science is also a hatred of these scientifically minded reformers:

When I see the spirit of liberty in action, I see a strong principle at work; and this, for a while, is all I can possibly know of it. The wild *gas*, the fixed air is plainly broke loose: but we ought to suspend our judgment until the first effervescence is a little subsided, till the liquour is cleared, and until we see something deeper than the agitation of a troubled frothy surface.

Employing an image of the carbonic acid – wild gas – which Priestley used to make soda water, Burke designs an image of liberty being dangerously agitated by 'sophisters, oeconomists and calculators' who adhere to a barbarous, 'mechanic' philosophy. These radical theorists act by the 'organic *moleculae* of a disbanded people'. They are drunken, violent, dangerously intelligent barbarians whose 'new-sprung modern light' Burke furiously rejects.

Punningly identifying political agitators with scientists who experimentally agitate gas and liquids in their laboratories, Burke indulges in a sprawling series of related images. He damns the 'confused jargon' of the Dissenters' 'Babylonian pulpits' and the activities of the French Revolutionary 'state surveyors', who by their violent haste and defiance of the laws of nature have blindly delivered themselves over 'to every projector and adventurer, to every alchymist and empiric'. His aim is to smear rational dissent, scientific theory and economic theory – everything geometrical, arithmetrical, abstract.

What Burke particularly detests is the feistiness of rational dissent, and it is notable that on several occasions he angrily, desperately, deliberately pits the unlovely adjective 'sluggish' against these modern activists. Thus property as an interest is 'sluggish, inert and timid', while we – Burke is ventriloquizing for the English – are supposed to be 'a dull sluggish race'. Our resistance to innovation, he later adds, is thanks to 'the cold sluggishness of our national character'. Yet who is less sluggish, less able to say 'we', than the immigrant Irishman Burke, whose agonized, passionate intellect must seek acceptance among English gentlemen by putting forward arguments against new scientific and political ideas?

His championing of a sluggish lack of ideas is linked to the famous passage in *Reflections* where the 'importunate chink' of a mere half-dozen grasshoppers is contrasted with the silent, cud-chewing thousands of 'great cattle reposed beneath the shadow of the British oak'. Sluggishness is therefore a quality that belongs to the national preference for the unwritten, the intuitive, the ancient and concrete as against the abstract and theoretical, but it is more than this.

Burke's use of the adjective is a challenge thrown back to dissenting science because the term 'sluggish' is rejected by Priestley as a quality that enforces a traditional and false distinction between body and spirit. Priestley opposes the view that matter has an evil tendency, arguing that this idea is the result of a dualism which divides reality into the 'sluggish body' and 'immortal spirit' (the term 'sluggish' was first used to describe matter in 1640). Hazlitt's poetics of bubbling motion follows Priestley's energetic materialism, as when he argues that Shakespeare's humour 'bubbles, sparkles and finds its way in all directions like a natural spring'. In 'The History of the Philosophical Doctrine Concerning the Origin of the Soul', which is included in *Disquisitions Relating to Matter and Spirit* (1777), Priestley argues that matter is not the '*inert* substance that it has been supposed to be, that *powers of attraction* or *repulsion* are necessary to its very being, and that no part of it appears to be *impenetrable* to other parts.' And he goes on to state that his argument tends

to remove the *odium* which has hitherto lain upon matter, from its supposed necessary property of solidity, inertness, or sluggishness; as from this circumstance only the *baseness* and *imperfection* which has been ascribed to it are derived. Since matter has, in fact, no properties but those of *attraction* and *repulsion*, it ought to rise in our esteem as making a nearer approach to the nature of spiritual and immaterial beings, as we have been taught to call those which are opposed to gross matter.

By insistently employing the adjective 'sluggish', Burke implicitly rejects Priestley's view both of society and matter: he is opposed to the release of energy in both social and material terms. And by using a term common in Priestley's writings – 'fixed air' – to characterize a dangerous idea of liberty, he denies the doctrine which true Whigs share with republican dissenting scientists. Where Priestley states that the 'complaint of the *evil tendency of matter* is a hackneyed topic of declamation among all the ancients, heathens and Christians', arguing that it is a mistake to divide reality into 'sluggish body' and the liberty of the 'immortal spirit', Burke asserts traditional scientific doctrine, in support of the monarchy.

Yet Burke himself is wild gas on the loose, and he rampages through the 1790s like a man who is at last free to indulge in a total verbal riot. The element of sheer release from all control cannot be underestimated – it whoops out from his exaggerations like the collective cry of a rushing mob. His prose style enacts a type of apocalyptic meltdown as it participates in the destructiveness it decries.

Burke's irrationalism was deeply influential, and it helped to shape the persecution which the Unitarians suffered in the 1790s. His romantic hatred of science had lasting effects, one of which was the view that Price and Priestley were foolish dreamers intent on destroying an ancient baggy state stuffed with tradition, spirit and communal sentiment. He did lasting damage to the Unitarian cause and, though it would be wrong to blame Burke and his followers for this, the fact that there is no definitive history of rational dissent in these islands needs to be registered as a glaring gap, an impoverishing absence.

Were such a history to be written, it would show how central rational dissent has been to the intellectual, scientific and political life of Britain and Ireland. Priestley's ideas met with an enthusiastic response in the North of Ireland, for example, where the republican activist Dr William Drennan drew on them as he formulated the idea of an independent Irish republic. Drennan married a member of the Revd William Hazlitt's Shropshire congregation, Sara Swanwick, who was the sister of Hazlitt's boyhood friend, Joseph Swanwick. In common with many Unitarians, Drennan was keenly interested in science (science and economic theory are two of the central pillars of Unitarian culture). On 28 March 1794 – after the destruction of Priestley's meeting house and his laboratory by a Church and King mob, and shortly before his emigration to North America – Drennan published an address entitled *The Society of United Irishmen of Dublin, To Joseph Priestley, L.L.D.* This pamphlet,

almost inaccessible today, is printed in full below. In it the language
– and attitudes – of experimental science in the 1790s are fused
intensely with the passionate rhetoric of liberty.

THE SOCIETY OF
UNITED IRISHMEN OF DUBLIN
TO JOSEPH PRIESTLEY, L.L.D.

SIR,

SUFFER a Society which has been calumniated as devoid of all
sense of religion, law, or morality, to sympathise with one, whom
calumny of a similar kind is about to drive from his native land, a
land which he has adorned and enlightened in almost every branch
of liberal literature and of useful philosophy. The emigration of
DOCTOR PRIESTLEY, will form a striking historical fact by which
alone, future ages will learn to estimate truely the temper of the
present times. Your departure will not only give evidence of the
injury which philosophy and literature have received in your
person, but will prove that accumulation of petty disquietudes,
which has robbed your life of its zest and enjoyment, for, at your
age, no one would willingly embark on such a voyage, and sure we
are, it was your wish and prayer to be buried in your native country,
which contains the dust of your old friends, SAVILE, PRICE, JEBB
and FOTHERGILL. But be chearful, Dear Sir, you are going to a
happier world – the world of WASHINGTON and FRANKLIN.

In idea, we accompany you. We stand near you while you are
setting sail. We watch your eyes that linger on the white cliffs, and
we hear the patriarchal blessing which your soul pours out on the
land of your nativity, the aspiration that ascends to God for its
peace, its freedom, and its prosperity. Again, do we participate in
your feelings on first beholding nature in her noblest scenes, and
grandest features; on finding man busied in rendering himself
worthy of nature, but more than all, on contemplating with philo-
sophic prescience, the coming period when those vast inland seas
shall be shadowed with sails, when the St Lawrence and Mississippi,
shall stretch forth their arms to embrace the continent in a great
circle of interior navigation; when the Pacific ocean shall pour into

the Atlantic; when man will become more precious than fine gold, and when his ambition shall be to subdue the elements, not to subjugate his fellow creatures, to make fire, water, earth and air obey his bidding, but to leave the pure ætherial mind, as the sole thing in nature, free and incoercible.

Happy indeed would it be were men in power to recollect *this* quality of the human mind. Suffer us to give them an example from a science of which you are a mighty master, that attempts to fix the element of mind only increases its activity, and that to calculate what may be, from what has been, is a very dangerous deceit. Were all the salt petre in India monopolized, this would only make chemical researches more ardent and successful. The chalky earths would be searched for it and nitre beds would be made in every cellar and every stable. Did not that prove sufficient the genius of chemistry would find in a new salt a substitute for nitre or a power superior to it.* It requires greater genius than Mr Pitt seems to possess, to know the wonderful resources of mind, when patriotism animates philosophy and all the arts and sciences are put under a state of requisition, when the attention of a whole scientific people is bent on multiplying the means and instruments of destruction, and when philosophy rises in a mass to drive on the wedge of war. A black powder has changed the military art and in a great degree the manners of mankind. Why may not the same science which produced it, produce another powder which inflamed under a certain compression, might impel the air so as to shake down the strongest towers and scatter destruction.

But you are going to a country where science is turned to better uses. Your change of place will give room for the matchless activity of your genius, and you will take a sublime pleasure in bestowing on Britain, the benefit of your future discoveries. As matter changes its form but not a particle is ever lost, so the principles of virtuous mind are equally imperishable, and your change of situation may even render truth more operative, knowledge more productive, and, in the event, liberty itself more universal. Wafted by the winds, or tost by the waves, the seed that is here thrown out as dead, there shoots up and flourishes. It is probable that emigration to America

* Mr Berthollet discovered that oxygenated muriatic gas, received in a ley of caustic pot ash, forms a chrystallizable neutral salt which detonates more strongly than nitre.

from the first settlement downward, has not only served the cause
of general liberty, but will eventually and circuitously serve it even
in Britain. What mighty events have arisen from that germ which
might once have been supposed to be lost for ever in the woods of
America, but thrown upon the bosom of nature, the breath of God
revived it, and the world has gathered its fruits. Even Ireland has
contributed her share to the liberties of America, and while purblind
statesmen were happy to get rid of the stubborn Presbyterians of
the North, they little thought they were serving a good cause in
another quarter. Yes! – the VOLUNTEERS of Ireland still live – They
live across the Atlantic. Let this idea animate us in our sufferings,
and may the pure principles and genuine lustre of the British Constitution, reflected from *their* coasts penetrate into our cells, and our
dungeons.

Farewell – great and good man! great by your mental powers, by
your multiplied literary labours, but greater still by those household
virtues which form the only solid security for public conduct, by
those mild and gentle qualities which far from being adverse to, are
most frequently attended with severe and inflexible patriotism, rising
like an oak above a modest mansion. Farewell, but before you go,
we beseech a portion of your parting prayer to the author of good,
for ARCHIBALD HAMILTON ROWAN, the pupil of JEBB, our
brother now suffering imprisonment, and for all those who have
suffered, and are about to suffer, in the same cause, the cause of
impartial, and adequate representation – the cause of the Constitution. Pray to the best of beings for MUIR, PALMER, SKIRVING,
MARGAROT, and GERALD, – who are now or will shortly be crossing,
like you, a bleak ocean but to a barbarous land. Pray that they may
be animated with that same spirit which in the days of their fathers,
triumphed at the stake, and shone in the midst of flames. Melancholy indeed it is that the mildest and most humane of all religions,
should have ever been so perverted as to hang, or burn men in
order to keep them of one faith. It is equally melancholy that the
most deservedly extolled of civil constitutions, should recur to similar modes of coercion, and that hanging and burning are not now
employed, principally because, measures apparently milder are considered as more effectual. Farewell! Soon may you embrace your
sons on the American shore, and WASHINGTON take you by the
hand, and the shade of FRANKLIN look down with calm delight,

on the first statesman of the age extending his protection to its first philosopher.

March, 28th 1794

Four years later the Society of United Irishmen, which Drennan helped to found, led an unsuccessful rebellion against the government. It was viciously suppressed and many of those stubborn Northern Presbyterians died on the battlefield or the scaffold. But Priestley's and Drennan's driving enthusiasm for new ideas, experiment and more equal social structures lived on within dissenting culture.

'It will not slice a pineapple': Babbage, Miracles and Machines

DORON SWADE

Charles Babbage (1791–1871) is honoured as the towering patriarch in the history of computing. The designs for his vast mechanical calculating engines rank as one of the startling intellectual achievements of the nineteenth century. We venerate his inventive genius and tell and retell the tale of how he failed to build a complete engine despite extravagant government funding, decades of design and development, and the social advantages of a well-heeled Victorian gentleman of science. Because of his failure to complete a physical engine, doubt has clouded his reputation. He was polymathically accomplished. But when it comes to his engine, was he more than an impractical dreamer? His failure to deliver an engine robbed him and history of the definitive proof. So there is an ambiguity in our historical perception of Babbage: reverence and ridicule. Reverence for the ingenuity and energy of his efforts; ridicule, touched perhaps with secret *schadenfreude*, for a man who so totally staked his personal and professional self-esteem on the demonstrable construction of a grand design.

The collapse in 1833 of a ten-year project to build Difference Engine No. 1 was the central trauma in Babbage's life. He wrote at different times in outrage, disbelief, disdain, protest and despair, as though unable to reconcile himself to the dismal outcome. But we should be wary as we weave this tale. There is a ready-made compartment in our historiographic sewing box for scientist as hero. Figures such as Pasteur, Semmelweis, Curie and Faraday swirl before us in images of bearded dedication, white coats, retort stands and glass flasks. Service to humanity shines in their eyes. Stiff poses frozen in the moment of greatness for posterity to revere. Statues, busts and portraits stare uncomprehendingly past us at some distant arrested vision. The hagiography of such figures touches us more if heroic achievement entailed obstacles, suffering, sacrifice or tragedy. We reserve a special garb for prescient scientists unrecognized in their day. These we clothe as visionaries out of time. And so it is with Babbage.

It is tempting to cast Babbage as the visionary progenitor of the modern computer, and there is much to encourage us so to do. In 1991, the bicentennial year of Babbage's birth, major computer companies sponsored a commemorative exhibition at the Science Museum as well as the construction from original designs of a Babbage calculating engine. *New Scientist, Scientific American* and the *Bulletin of the British Computer Society* ran cover feature articles on 'the architect of modern computing'. The Royal Mail included a 22p Babbage postage stamp in a set of four commemorating British scientific achievement. Babbage took his place in the philatelic hall of fame alongside Michael Faraday (electricity), whose bicentenary he shares, Frank Whittle (jet engine) and Robert Watson-Watt (radar). Industry openly identified with an historical figure of the last century, elevated in his bicentennial year to the status of national hero.

The modern computer industry is relatively new and self-consciously lacking in the kind of gravitas that accrues from having generationally distant founding fathers to oversee proceedings from mahogany-framed oil paintings in the boardroom. Not for this industry a proud plaque inscribed 'Founded in 1752'. In a culture in which precedent has vast power, the past is used to license the present. Adopting Babbage as ancestral patron suggests that the modern computer industry is not just an industrial parvenu hustling for markets but one whose roots lie deep in the past, and one that is therefore respectable, stable and secure. But the leapfrog from the modern industry back to Babbage blurs issues of developmental continuity.

Scientific training habituates us to see precedent as cause. *Post hoc, ergo propter hoc*. Babbage was the first to embody in his designs the principles of a general-purpose computational device. It is not a question of suggestive hints or the coded vagueness of Nostradamus.

Charles Babbage in his late fifties. Daguerreotype by Antoine Claudet,
c. 1849.

Essential logical features are described in explicit detail in hundreds of design drawings, and the development of his thinking is recorded in some twenty 'scribbling books'. The volume, range and detail of this work, which occupied him for decades, will shame anyone suspicious of his growing fame. Calling Babbage the first 'pioneer of the computer' is not a casual tribute. But because he was the first it is assumed that the modern electronic computer has Babbage as its patrilinear source. He is repeatedly referred to as the father, grandfather, forefather, great ancestor of the modern computer. The language of fatherhood serves to reinforce the notion of an unbroken line of descent. But the lineage of the modern computer is not as clear-cut as these genealogical tributes imply.

Our fascination with his failures may have a cultural dimension. The combination of eccentricity, genius and failure perhaps touches something deep in English culture. Genius aside, witness the phenomenon of Eddie 'The Eagle' Edwards, the first British ski jumper to participate in a winter Olympics. The myopic Eagle apparently landed on his head after a jump, the trajectory of which fell short of the nearest trailing competitor by a decent number of metres. He returned from Canada to a hero's welcome at Heathrow airport and enjoyed some celebrity thereafter. 'The Eagle lands amid whirlwind of adulation' announced *The Times* in March 1988 on his arrival from the Calgary Olympics. 'Britain's best bad skier is back' – this above a photograph showing Eddie besieged by the paparazzi. As recently as March 1994 we find 'Eagle on losing streak' – the famously unsuccessful British ski jumper was to present a new radio series about losers.

There are other temptations to lionize Babbage and to celebrate his tale. The most frequently quoted reason for Babbage's failure to realize any of his designs in physical form is that Victorian mechanical engineering was inadequate to the task. The thesis has a distinct appeal, at least at face value. We have a tendency to regard the technological past as crude. From the high ground of modern microelectronics it is tempting to portray Babbage as trapped in his world of cogs and levers, and to regard his designs as a product of visionary genius, but unrealizable in a mechanical medium. *We* succeeded where Babbage failed. Babbage had to wait for us, the children of the silicon age, to realize his dreams in physical form. We appropriate his genius by patronizing his failures. E. P. Thompson confected a delicious phrase. He warns of 'the enormous condescension of posterity' in our presumptions of superiority over superseded trades and crafts. The warning seems to have gone unheeded.

It is the ubiquity of computers and the way they have forced themselves on our attention that places a special value on the excavation of this specific past. We struggle in perplexity to absorb the computer into our assumptive worlds. We seek to locate in some familiar framework its pretensions to intelligence, its versatility and its paradoxical combination of plasticity of function and brittleness of design, that is to say, the ease with which its function can be changed through its software, set against its rigid intolerance of minor faults or slight inexactness. Were it not for the vast present-day importance of the computer, Babbage would not enjoy the stature of national hero. He would occupy instead a colourful footnote in the history of early Victorian science, and his wondrous designs would be no more than an elaborate curiosity.

The reinvention of the fundamental logical features of computers by electronic engineers of the 1940s, none of whom had more than scant knowledge of Babbage, equips us to recognize the extraordinary prescience of Babbage's work. After he had long since despaired of building his Analytical Engine he wrote prophetically:

If, unwarned by my example, any man shall undertake and shall succeed in really constructing an engine ... upon different principles or by simpler mechanical means, I have no fear of leaving my reputation in his charge, for he alone will be fully able to appreciate the nature of my efforts and the value of their results.

We can read this passage as a reassertion of his belief in the value of his work, or perhaps as a rhetorical declaration of confidence in the judgement of posterity, through which he sought some solace for the injuries and disappointments of his own self-perceived failures. It can also be read as an expression of his recognition that he had identified something fundamental, and even universal, about the systematic procedures of machine-like behaviour – ideas that were not formalized until the 1930s by the English mathematician Alan Turing.

While the direct indebtedness of the modern computer age to Babbage remains problematic there is no question that his engine designs represent an astonishing and original leap both in physical size and logical conception. Maurice Wilkes, a distinguished pioneer of modern electronic computers, led the postwar team at Cambridge that built the first usable electronic computer, the EDSAC. In 1971, the centennial year of Babbage's death, Wilkes published an article 'Babbage as a Computer Pioneer', which was the earliest authoritative evaluation of Babbage's contribution in modern times, and he has since described Babbage as possessing 'vision verging on genius'. It is therefore shocking to find Babbage accused by Wilkes, not of pioneering the modern computer age, but of delaying it. Wilkes argues that Babbage became associated with failure and that this discouraged others from advancing the cause of automatic computation. I have always taken this assertion as a rhetorical device to offset any tendency to lionize Babbage – an attempt as it were to bend the celebratory wire beyond its neutral point in the opposite direction to rebalance any tendency to hysterical excess. But evidence has since come to light of at least one instance in which Wilkes's allegation, however originally intended, is specifically and historically valid.

Thomas Fowler, a self-taught Devonshire printer and bookseller, devised an original calculating device. Fowler had neither social position nor means. He was moreover humble and far from the bustling London social and scientific scene. In 1841 he wrote with touching vulnerability to George Biddell Airy, Astronomer Royal and *de facto* scientific adviser to government, about his hopes for exhibiting his machine at a forthcoming meeting of the British Association for the Advancement of Science.

I have led a very retired life in this town without the advantage of any hints or assistance from anyone, and I should be lost amidst the crowd of learned and distinguished persons assembled at the meeting, without some kind friend to take me by the hand and protect me.

Fowler's machines were rendered in wood rather than metal. The machine was demonstrated in London and exhibited at King's College to some acclaim. Dignitaries of the mathematical establishment including Babbage, the astronomer Francis Baily and the logician Augustus De Morgan witnessed the device in action and confirmed its originality and efficiency. Fowler's machine differed in essential respects from Babbage's. Babbage's engines used the familiar decimal system with the numbers 0 through 9 each of which was represented by a discrete position of a rotating gear wheel. Fowler's machine on the other hand was more fully digital in that it used as its active element not rotating wheels but sliding rods which could occupy only one of three positions at any time. The advantage of reducing the number of distinct physical states is that parts can be made less precisely, and wooden devices by all accounts were practicable. In the light of Konrad Zuse's work in Germany on mechanical and electromechanical digital devices in the 1930s and 1940s, and the almost universal adoption of binary (two-state) digital techniques in the electronic computer age, Fowler's calculator was in certain respects vastly more promising than Babbage's. It is with unmistakable bitterness that Thomas Fowler's son wrote in a biographical notice following his father's death that 'the government of the day refused even to look at my father's machine, on the express ground that they had spent such large sums with no satisfactory result, on Babbage's "calculating engine"'. Bitterness is tempered by pathos when we learn that Fowler dictated details of his engine to his daughter on his deathbed. As with Babbage, the fate of innovation depended on vastly more than the promise or capability of the technology alone.

The Fowler episode has come to light only recently and was unknown to Wilkes in 1971. (I found the Airy–Fowler correspondence in 1993.) A few years ago I asked Wilkes what he had in mind in alleging that Babbage had delayed the development of computers. He was quite clear. Wilkes had known L. J. Comrie, an acknowledged authority on the calculation and production of mathematical tables, the selfsame task that first motivated Babbage to design calculating machines. Comrie spent eleven years at the Nautical Almanac

Office which annually published mathematical tables for navigation, and he served as its Superintendent from 1930 to 1936. Comrie is a linking figure between the failed nineteenth-century movement to mechanize calculation and the modern computer age. Comrie knew of Babbage's earlier attempts, and had interpreted the Babbage episode as a lesson not to undertake the construction of purpose-built machines for computation, but to wait for commercially developed devices which he could then adapt. His distinguished service producing almanacs and the founding of an independent scientific computing service in 1937 suggests that the lesson was not only sound but well learned.

Fowler's tale and the sorry saga of Babbage's failed attempts to construct a complete engine provide historical instances where science and human affairs intersect. But that very statement presupposes that science can be abstracted from human affairs at all. Jack Morrell, in an essay on the professionalization of science, observes that when William Whewell coined the term 'scientist' in 1833, this served to distinguish those exploring the material world from those concerned with literary, religious, moral and philosophical realms. The word 'scientist' had a separatist function from the start.

The culture of science encourages us in the notion that its content is value-free. Scientific knowledge is supposedly 'objective' in that it appears to belong not to the practitioner but to the object. If *you* measure the surface tension of water and *I* measure the surface tension of water then the result is apparently the same and does not depend on our subjective differences. The notion that science has some privileged access to certainty through its methods, and that the content of science is value-free, are implicit constructs of modern scientific training. The isolation of the internal content of science from human affairs is something we now take for granted. 'Hard' science, exemplified by physics, does not include humanity as its partial subject. A physicist or applied mathematician specializing in ballistics or kinematics is interested in my terminal velocity when falling from an aircraft under the influence of gravity, in my density or in my volume. He is interested in me in much the same way as he is interested in a stone. It is striking therefore to read of the fierce arguments that raged in the late 1820s when the young Darwin was a student of medicine at Edinburgh. Science and politics were not yet immiscibly distinct. Phrenology was used as an argument against hereditary privilege, patronage and preference. If talents were evident to all from bumps on the head then high birth or social connections could not presume to endow one automatically with superior ability. Merit and privilege were in the process of being uncoupled, and evidence from science seemed to belong as obviously to the argument as did assertions from other parts of contemporary thinking. Early evolutionary theory, with its notions of competition between species, was being used as a moral justification for capitalism – nature licensing a competitive free-for-all. Nowadays it seems to be different. Intellectuals, artists and novelists have recently begun to explore the connections between modern quantum mechanics and postmodernism. But the political charge of such debate seems a little damp.

The early Victorian élite embraced polymathy and inherited the assumption that all that was known or knowable could be mastered. Being a know-all was not yet presumptuous. The practical realization that the full gamut of scientific knowledge could no longer be marshalled by a single practitioner began to surface in Babbage's day as science began to fragment into specializations. With specialization came the curse of learned ignorance. Not only was science disintegrating as it flourished but the fault lines between science and religion became stressed as they began to compete for control over the accepted view of the world. Geology was producing evidence that the teachings of natural theology as to the age of the earth were untenable, and pre-Darwinian evolutionary theories were beginning to mount a challenge to the Book of Genesis.

Miracles posed particular difficulties for rational science. By definition miracles are events without evident cause. Discontinuities, singularities, and unpredictable phenomena defied the doctrines of causality espoused by science. Miraculous phenomena posed no such problems for religion. They were not only manifest proof of the omnipotence of the deity but very good PR. So science was under pressure to engage with the issue of miracles. It did so at times with incomprehensible leadenness. There was a growing fashion for statistics and Babbage joined in the elaborate attempt to deal with miraculous events using probability theory. The causally *outré* was brought into the grasp of statistics by construing the term 'miraculous' to mean 'improbable'. Babbage compared the mathematical probability of the occurrence of the miracle itself with the mathematical probability of the truth of multiple independent witness accounts, any of which may be faked or mistaken. The argument is not overpoweringly clear, though Dickens's Gradgrind would have been proud of such a spirited attempt to reduce wonder to number. More improbably we find Babbage again using probability calculus to compute the numerical probability of the Resurrection – this without a flicker of Swiftian absurdity. (He made it 200,000 million to one against.) Most of Babbage's deliberations on the nature of miracles appear in his portentously titled Ninth Bridgewater Treatise published in 1837. Eight official treatises were commissioned by the Royal Society under the terms of the will of the eighth Earl of Bridgewater who bequeathed the sum of £8,000 for the production of works 'On the Power, Wisdom, and Goodness of God, as manifested in the Creation'. Babbage did not receive an official commission and, without any evident embarrassment at this exclusion, he produced his supernumerary 'ninth' uninvited. The work is in part a response to William Whewell's official treatise 'Astronomy and General Physics considered with reference to Natural Theology', published in 1833. The formidable Whewell asserted that scientific and religious modes of thought were irreconcilable and that mathematicians and mechanists disqualified themselves *ex officio* from theological debate. This was anathema to Babbage, a devout rationalist for whom the highest truths were mathematical, and for whom empirical evidence provided the only sound basis for belief. Whewell's point that science was a discourse the validity of which was relativized to a particular narrow class of physical phenomenon was lost on Babbage, for whom rational science was coterminal with truth.

It is in the context of the widening gap between science and

religion that Babbage's Difference Engine offers an elegant example
of the role played by a physical artefact in the history of ideas. After
some ten years of design and manufacture Babbage still had nothing
in the way of a working engine. To prop up his flagging credibility
he instructed Joseph Clement, his engineer, to assemble a small
section of Difference Engine No. 1 from the thousands of loose parts
already completed. This, 'the finished portion of the unfinished
engine', completed in 1832, was displayed in the drawing room of
his house in Dorset Street, Marylebone. It was an object of marvel
at Babbage's Saturday soirées frequented by the social, literary and
intellectual élite of London, and where objects of scientific and artistic curiosity could be seen. Members of the privileged set were peti-
tioned by hopefuls to display their inventions and contrivances to
the toffs of science, literature and art. Not everyone was preoccupied
with the advancement of mind. The geologist Charles Lyell impor-
tuned Babbage to invite Colonel Codrington, briefly in town, to give
him an opportunity to meet Codrington's wife, who Lyell had
heard was 'very pretty'. Lyell urged Darwin, recently back from his
five-year adventure on the *Beagle*, to attend Babbage's West End
soirées where he would meet the fashionable intelligentsia and,
moreover, 'pretty women'. For the entertainment and instruction
of London's glitterati Babbage devised a demonstration using his
Difference Engine to reconcile the notions of rational order and
miraculous events.

He would set the machine to work to a simple rule. Each time he
operated the handle the numbers on the engraved figure wheels
would, for example, increment by two. So the sequence of numbers
read off by his enraptured guests would be 0, 2, 4, 6 ... After
many such repetitions the onlookers developed a fair degree of confidence as to the inevitability of the next number expected. After,
say, a hundred such repetitions the onlookers were witness to a
remarkable event. Without in any way interfering with the engine,
on the next operation of the handle the number leapt not by two
but, say, by 117. You see, says Babbage turning from the machine,
to you the onlookers this leap appears as a violation of law, that is
the law of incrementing by two. But I instructed the machine before

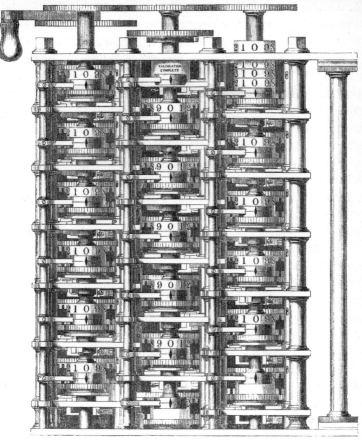

B. H. Babbage del.

Woodcut of the completed portion of Charles Babbage's Difference Engine No. 1 assembled in 1832. (Science Museum, London)

the demonstration so that after 100 repetitions it would add not two but 117. So for me, the programmer, the discontinuity was not a violation of law but the manifestation of a higher law, known to me but not to you. By analogy, concluded Babbage, miracles in nature are not violations of natural law but the manifestation of a higher law, God's law, as yet unknown. Apparent anomalies in nature were thus programmed discontinuities. For Babbage, God was a programmer. (Those in the software industry may be flattered to know this.) Control engineering also seems to have been part of God's education. Babbage writes of the deity revisiting the earth at some distant time to check for empirical deviations from the grand design which he (God) would then correct. The fact that such deviations might compromise divine perfection did not seem to bother Babbage. Empiricism embodied the highest ideal of human aspiration. We can see Whewell in his Cambridge college developing a sudden toothache.

The machine displayed by Babbage is arguably the most celebrated icon in the prehistory of computing. It was the first successful automatic calculator and symbolizes the start of the era of automatic computation. For the first time mathematical rule was successfully embodied in mechanism. No knowledge of its internal workings or of the mathematical principle upon which it was based was necessary for an operator to achieve useful results. By exerting physical effort (cranking a handle) one could arrive at results which up to that time could be achieved by mental exertion alone. The temptation to say that the machine was 'thinking', as Lady Byron (Ada Lovelace's mother) did, was irresistible. 'The marvellous pulp and fibre of the brain had been substituted by brass and iron, he [Babbage] had taught wheelwork to think.' So wrote Harry Wilmot Buxton, a younger contemporary of Babbage, after Babbage's death in 1871. As Simon Schaffer demonstrates, the metaphor of machine intelligence, if not a very developed view of its implications, is clearly in evidence. Babbage's Difference Engine was obviously not the first automatic machine of the industrial movement and takes its place alongside clocks, trains, textile machinery and a host of other devices and systems. But it is a landmark in respect of the human activity it replaced. In the case of textile machines or trains, the human activity they replaced was physical. The 1832 engine represents an ingression of machinery into psychology.

Babbage's mastery of mechanical design was matched by his political ineptitude. The exacting causality of machines and the uncompromising determinism of their design were features for which he had an inventive affinity unparalleled in his time. He had less affinity with the inconsistency, contingency and compromise of human affairs. He was, for one thing, a rotten publicist. During the decades of inspired design he published practically nothing of the detailed development on his calculating engines, preferring to rely on others to describe and promote his work. Touchy, proud and principled, he disdained to advertise. While demonstrating the engine to a group of personal friends at the 1862 London International Exhibition, he was twice heckled by members of the public about his notorious crusades against public street nuisances, organ grinders in particular. Offended, he kept clear of the engine thereafter, leaving further demonstrations to Wilmot Buxton and the engineer William Gravatt.

In 1842 he pressed for a meeting with the beleaguered Peel, in Prime Ministerial office just over a year. The year was perhaps the least auspicious of the century, dogged as it was with economic crises, industrial depression, starvation and violence. Peel was sick of the endless submissions from those hopeful of patronage, honours and financial awards. He sarcastically complained that he 'had not had a single application for office from anyone fit for it' and that being without a civil honour was becoming a rare distinction. He was also exhausted from the burdens of office and wrote in August to Julia, his wife, 'I am fagged to death.' Into his office bounds our hero. A man less proud, principled and aggrieved than Babbage would have done some homework, read the signs and prevailed upon Peel, who was publicly well disposed towards science and technology, for support. Not Babbage. Intent on discharging his sense of injustice, he gave voice to resentment and indignation at the failure of successive governments to honour their obligations to him. He insisted on his right to reward, laboured his sense of injury and accused those from whom Peel had sought advice of malicious intent. Peel was unyielding. Babbage was furious: 'If these are your views, I wish you good morning.' The meeting was the last opportunity to secure renewed government support for the engine project, then stalled for some ten years. With his exit went any prospect of completing his machine.

The politics of scientific organizations were equally opaque to him. Babbage did not appreciate or acknowledge the need to secure allegiance and enlist support from members of scientific coteries and clubs. He threatened in 1838 to resign from his co-trusteeship of the British Association in protest over the re-election of the geologist Roderick Impey Murchison (a co-trustee) to the post of General Secretary. Murchison's sin was to have urged the Presidency of the Association on Babbage, but then later to recommend the astronomer John Herschel for the appointment without informing either of his actions. The unsought competition, growled Babbage, had imperilled a quarter-century of friendship with Herschel. Babbage assumed that as a trustee his criticism of an official would lead to Murchison's dismissal or at least his censure. The upshot was not what he expected. Murchison continued in office; Babbage was deprived of his trusteeship with no notice. He had been misled by self-importance and had misread the domestic politics of institutional power.

He had a run-in with the Royal Astronomical Society in 1847 over the award of its annual medal. The discovery of the planet Neptune was a matter of great public and scientific interest. John Adams at Cambridge had concluded after laborious calculation that the irregularity in the orbit of Uranus was consistent with the existence of an unknown planet, and he estimated its size and position. George Biddell Airy, Astronomer Royal, with whom Adams had deposited his results, elected not to verify the prediction by telescopic observation – an omission for which he was later vilified. Meanwhile the French astronomer Urbain Jean Joseph Le Verrier had independently predicted the new planet and sent his findings to several astronomers who had powerful telescopes. The planet was found by Galle in Berlin in 1846 following a memoir from Le Verrier. The issue of precedence in the discovery of Neptune was one of bitter recrimination and considerable national pride. The Royal Astronomical Society decided not to award its annual medal to Le Verrier. Babbage wrote in protest suggesting that both Le Verrier and Adams be honoured, Le Verrier by receiving the 1846 annual award, and Adams a special medal dated 1847. Babbage had been one of the founders of the Astronomical Society, established in 1820, though by 1847 his contact with its administrative affairs had long since lapsed, and he was insulted when his letter recommending joint awards was not read at the extraordinary general meeting convened to resolve the issue of the Neptune medals. He complained of the 'great discourtesy' of being so snubbed and wrote to *The Times* protesting the circumstances. His self-perception as an honoured elder statesman of the Society was at variance with the political agendas of incumbent officers. For Babbage, principle and pride were all. The play *Jacobowsky and the Colonel* by Werfel and Behrman centres on the uneasy wartime alliance, forced by the advancing German army, between a grandiose Polish colonel full of the dignity of office, high-minded principle and impractical grand designs, and a humble worldly-wise tailor inured to compromise. The exacting logic of mathematics and machines was Babbage's genius and his forte. But he needed a Jacobowsky.

Babbage was as well known in his lifetime for his failure to build his engines as he was for his sarcastic public diatribes against the scientific establishment. His *Decline of Science in England* published in 1830 was a shocking breach of the conventions of the day. He spared his readers the courtesies of gentle chiding or a discreet word in an appropriate ear. *Decline* was a broadside of outrage and insult. He accused the Royal Society of corruption and dereliction of duty in its service to science. He cited his quarry by name and openly impugned the personal and professional probity of each. His accusations included allegations of fiddling the minutes, misappropriating funds and rigging awards. As ever, Babbage argued more to protest than to persuade. John Herschel, model ambassador for science, saw a prepublication draft. Herschel was a close friend and told Babbage that had he been within reach he would have given

Babbage a 'good slap in the face' for so unworthy a piece. Babbage seemed to think that being right entitled him to be rude. In the early 1820s the Royal Society, the object of his denunciations, had been instrumental in securing government funding for his engine project. He had a knack, did Babbage, of alienating those whose support he most needed through a curious insensitivity to their position. Despite his spirited attacks on the Royal Society he never resigned his Fellowship.

Babbage's main effort to construct his Difference Engine collapsed in 1833. The circumstances were complex: an unresolved dispute with his engineer over financial compensation for moving the work from Lambeth to a special fireproof workshop in the grounds of Babbage's house, lack of credible progress and recurrent financial difficulties. Nowhere in this tangled tale is the technical impossibility of the engine cited as a reason for the cessation of work. The question that began to tantalize a few Babbage scholars was whether the complex of circumstances surrounding the abandonment of the project concealed the technical or logical impossibility of the design. Simply put, if Babbage's engine had been built, would it have worked? The question as to whether Babbage was an impractical dreamer or a designer of the highest calibre remained unanswered for over 150 years. A fortuitous cocktail of circumstances combined to allow a unique experiment that went a long way towards answering this question. The bicentenary in 1991 of Babbage's birth provided the public justification, and a supportive consortium of computer companies provided the means. Based on the Science Museum's comprehensive holdings of Babbage's mechanical relics and archival sources we built a Babbage Difference Engine to original designs. Babbage's Difference Engine No. 2 stands eleven feet long and seven feet high. It consists of 4,000 parts made of bronze, cast iron and steel and weighs just under three tonnes. The engine functions impeccably and calculates to thirty-one figures of accuracy. Its completion represents the culmination of a six-year project to raise funds, interpret the original design drawings, specify and manufacture parts, assemble and adjust. The saga of the construction is one worthy of Babbage himself: problematic funding,

technical challenges, the sudden liquidation of the company contracted to produce the parts and a desperate sprint to the post office
to mail specifications by closing time to meet contractual deadlines.
The engine performed its first full calculation in November 1991,
just over a month before the 200th anniversary of Babbage's birth.
The completion of the engine lays to rest an anguished episode in
the prehistory of computing.

Babbage envied the patronage of science in foreign countries,
France and Prussia in particular. His xenophilia was complemented
by indignation at the lack of support for English science, viewed by
many as being in comparative decline. He wrote in 1852 bemoaning the lack of entrepreneurial spirit in England and the habit of
perversely rejecting a device useful for one purpose because of its
deficiencies in another.

Propose to an Englishman any principle, or any instrument, however admirable, and you will observe that the whole effort of the English mind is directed
to find a difficulty, a defect, or an impossibility in it. If you speak to him of a
machine for peeling a potato, he will pronounce it impossible: if you peel a
potato with it before his eyes, he will declare it useless, because it will not
slice a pineapple.

The engine completed at the Science Museum is a sumptuous
piece of engineering sculpture. One of its many striking features is a
set of steel arms arranged in a series of vertical helices which rotate
like whirling scythes. Babbage should take comfort. There is little
doubt that while the engine is calculating with unerring accuracy,
a pineapple, held up to the mechanism, would be well and truly
sliced.

Difference Engine No.2: designed by Charles Babbage between 1847 and 1849, completed at the Science Museum, London, 1991. Standing: Reg Crick. Seated: Barrie Holloway. (Tony Cooper)

Babbage's Dancer and the Impresarios of Mechanism

SIMON SCHAFFER

They needed a calculator, but a dancer got the job.
The Marriage of Figaro, 1784

In the steam-punk metropolis of William Gibson and Bruce Sterling's *The Difference Engine* (1988), the sickly Keats runs a cinema, Disraeli is a gossip journalist unwillingly converted to using a keyboard, and fashionable geologists visit the Burlington Arcade to buy pricey mechanical trinkets, 'outstanding pieces of British precision crafts-manship'. Above them looms Lord Babbage, his original calculating engines already outdated, his scheme for life peerages on merit become part of everyday politics. Babbage's dreams doubtless de-serve this treatment from the apostles of cyberfiction – he touted his schemes in pamphlets and exhibitions all over early nineteenth-century London. It was a city apparently obsessed by displays of cunning engines, enthusiastic in its desire to be knowingly deceived by the outward appearance of machine intelligence, and Babbage heroically exploited the obsession in his lifelong campaign for the rationalization of the world.

The enterprise of the calculating engines was certainly dependent on the city's workshops, stocked with lathes, clamps and ingenious apprentices, and on government offices, stocked with ledgers, blue books and officious clerks – a heady mixture of Bleeding Heart Yard and the Circumlocution Office. But, as Gibson and Sterling see so acutely, it was also tangled up with the culture of the West End, of brightly lit shops and showrooms, of front-of-house hucksters and backroom impresarios. Put the Difference Engine in its proper place, perched uneasily between Babbage's drawing room in wealthy Marylebone, the Treasury chambers in Whitehall and the machine shops over the river in Lambeth, but at least as familiar in the arcades round Piccadilly and the squares of Mayfair, where automata and clockwork, new electromagnetic machines and exotic beasts were all put on show.

It was in the plush of the arcades that Babbage, barely eight years old, first saw an automaton. Some time around 1800 his mother took him to visit the Mechanical Museum run by the master designer John Merlin in Prince's Street, just between Hanover Square and Oxford Street. A Liègeois in his mid-sixties, working in London for four decades, Merlin was one of the best-known metropolitan mechanics, deviser of new harpsichords and clocks, entrepreneur of mathematical instruments and wondrous machines. His reputation even rivalled that of Vaucanson, the pre-eminent eighteenth-century designer of courtly automata. As he rose through fashionable society, Merlin hung out with the musical Burney family, figured largely as an amusing and eccentric table companion, and 'a very ingenious mechanic', in Fanny Burney's voluminous diaries, sat for Gainsborough and used his mechanical skills to devise increasingly remarkable costumes for the innumerable masquerades then charming London's pleasure seekers. To help publicize his inventions, Merlin appeared at the Pantheon or at Ranelagh dressed as the Goddess Fortune, equipped with a specially designed wheel or his own newfangled roller skates, as a barmaid with her own drink stall, or even as an electrotherapeutic physician, shocking the dancers as he moved among them.

Merlin ingeniously prowled the borderlands of showmanship and engineering. He won prestigious finance from the backers of Boulton and Watt's new steam engines. He opened his Mechanical Museum in Hanover Square in the 1780s. For a couple of shillings visitors could see a model Turk chewing artificial stones, play with a gambling machine, see perpetual-motion clocks and mobile bird cages, listen to music boxes and try the virtues of Merlin's chair for sufferers from gout. After unsuccessfully launching a plan for a 'Necromancic Cave', featuring infernal mobiles and a fully mechanized concert in the Cave of Apollo, he began opening in the evenings, charged his visitors a shilling a time for tea and coffee, and tried to pull in 'young amateurs of mechanism'.

Babbage was one of them. Merlin took the young Devonshire schoolboy upstairs to his backstage workshop to show him some more exotic delights. 'There were two uncovered female figures of silver, about twelve inches high.' The first automaton was relatively banal, though 'singularly graceful', one of Merlin's well-known stock of figures 'in brass and clockwork, so as to perform almost every motion and inclination of the human body, viz. the head, the breasts, the neck, the arms, the fingers, the legs &c. even to the motion of the eyelids, and the lifting up of the hands and fingers to the face'. Babbage remembered that 'she used an eye-glass occasionally and bowed frequently as if recognizing her acquaintances'. Good manners, it seemed, could easily be mechanized. But it was the other automaton which stayed in Babbage's mind, 'an admirable *danseuse*, with a bird on the forefinger of her right hand, which wagged its tail, flapped its wings and opened its beak'. Babbage was completely seduced. 'The lady attitudinized in a most fascinating manner. Her eyes were full of imagination, and irresistible.' 'At Merlin's you meet with delight', ran a contemporary ballad, and this intriguing mixture of private delight and public ingenuity remained a powerful theme of the world of automata and thinking machines in which Babbage later plied his own trade.

Merlin died in 1803, and much of his Hanover Square stock was sold to Thomas Weeks, a rival 'performer and machinist' who had just opened his own museum on the corner of Tichborne and Great Windmill Streets near the Haymarket. The *danseuse* went too. The show cost half a crown, in a room over one hundred feet long, lined in blue satin, with 'a variety of figures inert, active, separate, combined, emblematic and allegorical, on the principles of mechanism, being the most exact imitation of nature'. Like Merlin, Weeks also tried to attract invalids, emphasizing his inventions of weighing machines and bedsteads for the halt and the lame. There were musical clocks and self-opening umbrellas, and, especially, a tarantula spider made of steel, that comes independently out of a box, and runs backwards and forwards on the table, stretches out and draws in its paws, as if at will. This singular automaton that has no other power of action than the mechanism contained in its body, must fix the attention of the curious.

Once again, seduction was an indispensable accompaniment of the trade in automata. One of the most famous automata of the early nineteenth century, a 'Musical Lady', was originally brought to London in 1776 by the great Swiss horologist Jaquet-Droz. His London agent Henri Maillardet put her on show after the turn of the century at the Great Promenade Room in Spring Gardens behind Whitehall: 'the animated and surprizing Motion of the Eye aided by the most eloquent gesture, are heightened to admiration in contemplating the wonderful powers of Mechanism which produce at the same time the actual appearance of Respiration'. The accomplished lady's eyes really moved, her breast heaved. 'She is apparently agitated', a contemporary remarked, 'with an anxiety and diffidence not always felt in real life.' Such shows often turned to titillating effect modish materialist philosophies which, in the wake of enlightened theories of sensibility and mesmeric strategies for restoring health, sought to mechanize the passions, and especially those of women. Maillardet's advertisements put love on sale:

> If the Poet speaks truth that says Music has charms
> Who can view this Fair Object without Love's alarms
> Yet beware ye fond Youths vain the Transports ye feel
> Those Smiles but deceive you, her Heart's made of steel
> For tho' pure as a Vestal her price may be found
> And who will may have her for Five Thousand Pounds.

The neat connection between passion, exoticism, mechanism and money permeated the showrooms. Since the 1760s, London designers, especially Merlin and his erstwhile employer James Cox, had built extraordinary automata for the East India Company's China trade, opened shops in Canton where mandarins could acquire mechanical clocks, mobile elephants and automatic tigers, and thus oiled the wheels of the booming tea business. Maillardet and his partners joined in the market. But this lucrative eastern commerce languished as, after American Independence and a huge reduction in the tea duty, the British entrepreneurs found it ever harder to balance imports of the precious leaf. Bengal opium and Indian cotton were now used to help pay for Chinese tea, and successive delegations to the Chinese imperial court failed to impose what they tended to see as rational economic relations. Cox's firm went broke, and, while Weeks never quite managed to revive it, he ruthlessly exploited the appealing orientalist gloss it gave his Haymarket show. The word 'factory', it is worth remembering, was used for company storehouses in the Indies before it was used to describe workshops back in Europe, and the automata shifted between both these worlds. According to Weeks's advertisements for his machines,

these magnificent specimens which constitute almost all the labour of a long life, and were all executed by one individual, were originally intended as presents for the east, they have, indeed, all the gorgeous splendour, so admired there, and we can fancy the absorbing admiration they would create in the harems of eastern monarchs, where their indolent hours must be agreeably relieved by these splendid baubles, which however are so constructed as to combine in almost every instance some object of utility.

The slippery move between images of languorous oriental baubles and honest utilitarian labour defines the significantly ambiguous place the automata occupied in a metropolis equally impressed by the mechanical ingenuity, excess wealth and eroticized luxury which all marked its new worldwide imperium. This was an apt stage for automatic Turks, mechanical elephants and clockwork women.

The silver dancer never went on show at Weeks's Museum, but stayed neglected in an upstairs attic. Blocked from the Chinese trade, and failing to win London audiences, Weeks's Museum closed and its nonagenarian owner died in 1834. By now Babbage was an engineer and entrepreneur in his own right, the heir to a fortune of over £100,000 from his banker father. Throughout 1834 he was in the toils of a disastrous dispute with his master machinist Joseph Clement, a fight which soon ended with the abandonment of the

Difference Engine project. At the start of the year, he presciently commissioned two demonstration models of the Difference Engine from the instrument maker Francis Watkins, who also supplied electromagnetic and mechanical equipment to the new Adelaide Gallery of Practical Science, an exhibition of newfangled steam guns, clockwork model steamboats and telegraphic devices, in the Lowther Arcade just off the Strand and round the corner from Weeks's old showrooms. And in the midst of these machine plans and troubles, Babbage also took the time to visit Weeks's auction and buy, for £35, the long-lost silver lady. He painstakingly restored the automaton and put her on a glass pedestal in his Marylebone salon in the room next to the unfinished portion of the first Difference Engine.

What was proper to a machinist's storeroom was slightly *risqué* in the drawing room of a gentleman of science – the naked dancer needed a dress. Though he commissioned a new robe from local dressmakers, Babbage initially made do with a few strips of pink and green Chinese crêpe, a turban, a wrap and 'a pair of small pink satin slippers, on each of which I fixed a single silver spangle'. She was a hit, drew amused if slightly off-colour jokes from his visitors and provided Babbage with the chance to teach a portentous moral about the decline of the industrial spirit in England. 'A gay but by no means unintellectual crowd' of English guests could all too easily be entertained by the dancer's 'fascinating and graceful movements'. Only sterner Dutch and American inquirers would bother to visit the Difference Engine next door. Babbage ever after used the divergence to teach his audience about the sinister contrast between foreign seriousness and domestic triviality, between the easy charms of the silver dancer and the demanding challenges of the calculating engine.

Babbage worked hard to make, then exploit, this distinction between catchpenny and serious machines. He expostulated noisily and persistently against music machines, organ grinders and steam engines and published a long pamphlet, *Street Nuisances* (1864), describing the persecution he'd suffered in his once peaceful Marylebone home: 'the neighbourhood became changed: coffee-shops, beer-shops, and lodging houses filled the adjacent small streets. The character of the new population may be inferred from the taste they exhibit for the noisiest and most discordant music.' Eventually 'Babbage's Act' against street music became law: 'a grinder went away from before my house at the first word', reported Babbage's friend the London mathematics professor Augustus De Morgan. And the barrier between popular machines and scientific ones had more than merely the advantage of domestic tranquillity. The story of the silver dancer was partly designed to help contrast the appeal of fashion with the demands of manufacture. In the midst of his tortuous negotiations of 1834 about funding a new Analytical Engine, Babbage told the Duke of Wellington that the switch from the older difference-based design to the new mechanism was not to be damned as modish novelty.

The fact of a *new* superseding an *old* machine in a very few years, is one of constant occurrence in our manufactories, and instances might be pointed out in which the advance of invention has been so rapid, and the demand for machinery so great, that half-finished machines have been thrown aside as useless before their completion.

This was scarcely likely to mollify a penny-pinching administration, but throughout Babbage's career he felt it necessary to explain to what he saw as an irredeemably puerile public how to spot the difference between the engines which could make them rich and intelligent and those which deluded them into the gaudy fantasies of tricksy parlour games and theatrical delights. The inquisitive polymath Sophia Frend, later De Morgan's wife, recalled that most of the audience for Babbage's engine 'gazed at the working of this beautiful instrument with the sort of expression, and I dare say the sort of feeling, that some savages are said to have shown on first seeing a looking-glass or hearing a gun.'

The obscure objects of desire embodied in the automata were never self-evidently distinct from any of Babbage's projects. For example, like the automata of Cox, Merlin and Weeks, the Difference Engine apparently was also an object of fascination to the Chinese, and one visitor from China asked Babbage whether it could be reduced to pocket size. Babbage replied that 'he might safely assure his friends in the celestial empire that it was in every sense of the word an *out-of-pocket* machine'. Indeed, in the later 1840s, when all his engine schemes had run into the sand, he cast about for new ways of raising money to revive them, including writing novels, but was dissuaded because he was told he'd surely lose money on fiction. One such entirely abortive scheme involved designing an automaton 'to play a game of purely intellectual skill successfully'. This was at least partly an attempt to assert the very possibility of building an automatic games player. Babbage knew, at least at second-hand, just how seductive gambling could be – his close friend Ada Lovelace, Gibson and Sterling's dark lady of the Epsom motor races, lost more than £3,000 on the horses during the later 1840s. 'Making a book seems to me to be living on the brink of a precipice', she was told by her raffish gambling associate Richard Ford in early 1851.

Babbage's attention turned to the prospects of a games machine. In a brief memorandum, he demonstrated that if an automaton made the right first move in a game of pure skill with a finite number of possible moves at each stage, the machine could always win. Such a device, he reckoned, must possess just those faculties of *memory* and *foresight* which he always claimed were the distinctive features of his Analytical Engine, the features which made it *intelligent*. So Babbage began to design an automaton which could win at noughts and crosses, planning to dress it up 'with such attractive circumstances that a very popular and profitable exhibition might be produced'. All his memories of Merlin, Weeks and the Regency world of mechanical wonders came into play. As he reminisced in his 1864 autobiography,

I imagined that the machine might consist of the figures of two children playing against each other, accompanied by a lamb and a cock. That the child who won the game might clap his hands whilst the cock was crowing, after which, that the child who was beaten might cry and wring his hands whilst the lamb began bleating.

But there was, of course, a hitch. One point of his games machine was to raise money for the more portentous Analytical Engine, and Babbage soon discovered that though 'every mamma and some few pappas who heard of it would doubtless take their children to so singular and interesting a sight', and though he could try putting three shows on at once, nevertheless the mid-Victorian public simply weren't interested any more. 'The most profitable exhibition which had occurred for many years', Babbage moaned, 'was that of the little dwarf, General Tom Thumb', Phineas Barnum's famous midget money spinner, displayed in 1844 before gawping London audiences at the selfsame Adelaide Gallery where a decade earlier the Difference Engine models, steam guns and electromagnetic engines had drawn large audiences. According to London journalists the Adelaide 'with its chemical lectures and electrical machines' had by the later 1840s 'changed its guise, and in lieu of philosophical experiments we have the gay quadrille and the bewildering polka'. So, however apparently distinct, the fate of the automata shows and the calculating engines was remarkably similar, as metropolitan fashion switched away from the machines that could simulate human motions and emotions to the high life where the genteel tried these activities out for themselves.

Ultimately, Babbage's Difference Engine suffered more or less the same end as a whole range of Victorian automata, ending its days as a museum piece. In 1842, when the Government finally abandoned the project, Babbage told them it should be carefully preserved and 'placed where the public can see it, for example, the British Museum'. In the event, in January 1843 it was put behind glass in the very middle of the new museum at King's College, London, alongside a vast collection of memorabilia and eighteenth-century scientific instruments made for George III. For two decades, as Babbage bitterly remarked, 'it is remarkable that during that long period no person should have studied its structure'. The Engine was briefly brought out for the London International Exhibition of 1862 and put in what Babbage called 'a small hole, closed in and dark', where scarcely anyone could see it and, so he reckoned, it would have needed about eight hundred square feet of wall space to lay out all the diagrams required to explain its principles. Such exhibitions were increasingly devoted not to machines but to their products and rapidly became the Victorians' favourite sites for display of mechanically produced commodities. Babbage was told that he could have no more space for his calculating engines because of the room required for an appealing display of children's toys. Once again, he reflected, the British had revealed themselves to be more interested in entertaining tricks than thoughtful engineering.

This dismal fate was scarcely the sole link between the machine shows and Babbage's engines. There was an even more intriguing one, since all these devices neatly captured the puzzle of mechanical passion, of the possibility that artifices could think and feel. In his essay on the games machine, interpolated in a chapter of his autobiography entitled 'The Author's Further Contributions to Human Knowledge', Babbage made the point in his characteristically pithy way. He asked his friends 'whether they thought it required human reason to play games of skill. The almost constant answer was in the affirmative. Some supported this view of the case by observing, that if it were otherwise, then an automaton could play such games.' Babbage set out to show that an automaton could do just this. For example, there would always be occasions when the automaton was faced with two equally good moves. Then Babbage would program it so that a random number could direct the machine's decision between them. 'An enquiring spectator who observed the games played by the automaton might watch a long time before he discovered the principle upon which it acted.' In combination with his celebrated principles of foresight and memory, this principle of random moves programmed in advance governed Babbage's stories about machine intelligence.

From the 1830s, his favourite party trick, for those who got bored watching the silver dancer, was to program his Difference Engine to print out a very long series of unchanging numbers and then suddenly switch to a new series, on the basis described in Doron Swade's essay. Was this not exactly like a miracle? he would ask his guests. In summer 1833 Ada and her mother, Lady Byron, the Princess of Parallelograms, saw the trick. Lady Byron described the encounter to William King, a rather conventionally pious Cantab she had persuaded to act informally as mathematics teacher for Ada. 'We both went to see the *thinking* machine (for so it seems).

The Machine could go on counting regularly 1,2,3,4 &c – to 10,000 – and then pursue its calculation according to a new ratio. There was a sublimity in the views thus opened of the ultimate results of intellectual power.' The miraculous counting game was obviously a crowd-pleaser, the implications of the machine's discontinuous outputs were allegedly clear and radical, and at least one of Babbage's guests, Charles Darwin, soon picked up the hint. Darwin saw that if apparently inexplicable discontinuities could really be the result of a system of mechanical laws laid down in advance, then here was a useful analogue of the way new species could emerge entirely through natural law. Indeed, for Babbage and his allies, this was turned into a definition of what made machines intelligent. They could foresee, they could remember, and they could switch their behaviour in ways which seemed random but were really determined.

And in the epoch of the new factory system and Chartist strikes, this was also just how economic journalists lauded the new machines of automatic industry. In his oleaginous work of industrial reportage, *A Tour in the Manufacturing Districts of Lancashire* (1842), the free trader William Cooke Taylor described a Manchester spinning mule which

recedes and then returns so gracefully that I was almost going to say the effect was picturesque. I can assure you that the brightness of the machinery, which looks like steel, and the regularity of its motions, produce a *tout ensemble* which has a novel and striking effect. It seems to me that the machines can do everything but speak.

These machines had been developed in Lancashire, following strikes in the cotton factories, to give employers more control over the production process. Cooke Taylor piously observed that in these factories 'the human agents work with all the exactness of machinery. So strange a combination of perfect despotism with perfect freedom never before existed, and to have produced such a state is one of the noblest triumphs of morality and intelligence.' There was no mistaking the moral that the intelligence belonged to the system, not the operatives. In his notoriously eulogistic *Philosophy of*

Manufactures (1835), published little more than a decade after Mary Shelley's novel, the Scottish science writer Andrew Ure lapsed significantly into the imagery of her *Frankenstein, Or The Modern Prometheus* to describe the new spinning mule as 'the *Iron Man* sprung out of the hands of our modern Prometheus at the bidding of Minerva – a creation destined to restore order among the industrious classes'.

So machine intelligence was a central theme of the politics of manufacture, just as it was being worked out in the London show-rooms and Babbage's workshops. In his long-drawn-out contests with the engineers of the Lambeth machine shops, the dominant theme was precisely the kind of intellectual property represented by the calculating engines and the skills required for their construction. Babbage might try to keep street organs and noisy proles away from his door, but he wanted the calculating engines indissolubly linked with his property, and even tried shifting the whole engine works from Lambeth to his own backyard. He told Wellington in 1834 that his ownership of the engines was complete, 'for they are the absolute creations of my own mind'. But it had been a tradition of the machinists that all their tools belonged to the workmen, not the customers or masters, and it was thus an extremely sensitive issue as to which aspects of the calculating engines' enterprise counted as tools, and which as finished work. It was not at all obvious to a master machinist like Joseph Clement, the designer of remarkable new planing machines and facing lathes, that Babbage's mind was the unique source of these engines' value. The intelligence they embodied, therefore, was a prize contested by engineers, design-ers, proprietors and financiers, and intelligence's place was a major aspect of the political geography of the industrial system.

Babbage's most successful publication of the 1830s, a thorough survey of this geography, charted in great detail the ways in which mechanization automated the production process and insisted that the division of labour could be applied to mental just as much as to mechanical operations. Copying machines were one of his principal themes, and he explained them by describing such automata as the Prosopographus and the Corinthian Maid, machines shown in the

Strand in the early 1830s which could apparently copy the likeness of any sitter. Babbage explained that such shows really relied on a concealed camera lucida where a backstage assistant using a pentagraph linked to the automaton's own hand could quickly produce a reasonably accurate portrait of the customer. Here intelligence turned out to be the result of concealed skill in alliance with ingenious mechanization. In all these places, indeed, the puzzle of thinking engines was wrapped up with the problem of selective vision. Cooke Taylor aestheticized the cotton factory so that its intelligence seemed vested in the machinery, not the labour force. Ure saw the machines as the immediate intellectual offspring of the manufacturers, just as his idol Charles Babbage claimed that the Difference Engine was the unique product of his own mind. Clement saw his own workshop as a place of intelligent skill and so refused to move his workmen and their tools to Marylebone, where they would be under Babbage's immediate gaze. There Babbage's calculating engines looked miraculously prescient because of his partygoers' ignorance of their original programs. And, inevitably, the point of the West End automata was to mimic the actions of mind by concealing the springs of their artful design. To see such devices as intelligent, it was necessary to ignore, or conceal, or divert one's gaze from the machinations that drove them and the human skill on which they all depended.

Babbage's dancer was never just a gaudy trick. She was rather an alluring emblem of the aestheticized gaze of the impresarios of intelligence. The attribution of intelligence and reason to any machine depended on the perspective of the machine's audience and on the visibility of the labour on which its performance relied. Babbage's friends told him that to play a game of skill needed human reason and so denied any automaton could do it unaided. They would always suspect that any automaton which could apparently play chess must somehow be accompanied by a rational human temporarily hidden from view. Babbage and his critics both had one spectacular and timely precedent, a notorious automaton chess player, first shown in London in 1783–4 on Savile Row, and then again from 1818 at the showroom in Spring Gardens, where the charming Musical Lady also found her home. This chess player was built at the end of the 1760s by an aristocratic Slovak engineer, Wolfgang von Kempelen, as an entertainment for Maria Theresa, and its subsequent career took it from central European court society to the more vulgar milieux of French and English showrooms.

Von Kempelen himself never took the device very seriously, and frankly confessed it relied on a blatant trick. He temporarily dismantled it in 1773. Before then, guests at his house in Bratislava, just downriver from the Habsburg capital, were shown upstairs through his workshop, stocked with tools and unfinished projects for steam engines, perpetual clocks and especially his favoured scheme for a speaking machine, into his study, decorated with antiques, curiosities and prints. There, in the middle of the room, stood a large cabinet running on castors, and behind it an impressive full-scale model of a seated Turk smoking a pipe. On top of the cabinet was screwed a chessboard, the object of the Turk's fixed attention. Von Kempelen would open both the front and the back of the cabinet, revealing an extraordinarily complex array of gearwheels, barrels and pulleys. The custom was to shine a candle into the cabinet to show that nothing could possibly be hidden, and the Turk's torso and legs would also be stripped bare. *Inanimate Reason* (1784), the significantly titled publicity sheet for von Kempelen's machine, reported that 'you see at one and the same time, the naked Automaton, with his garments tucked up, the drawer and all the doors of the cupboard open'. Then, after von Kempelen had wound up the automaton, giving it enough power to run for about a dozen moves, the games would begin, the Turk gracefully moving pieces with his left hand, nodding his head when giving check, tapping the table and replacing any piece if a false move was made by his opponent, and bowing to the spectators when the game ended, almost always with the Turk's triumphant victory. The ritual of open display and brilliant chess never varied. 'Never before did any mere mechanical figure unite the power of moving itself in different directions as circumstances unforeseen and depending on the will of any person present might require', and nowhere else in Europe did the relation between intelligence, mechanism and concealment become such a matter of public interest.

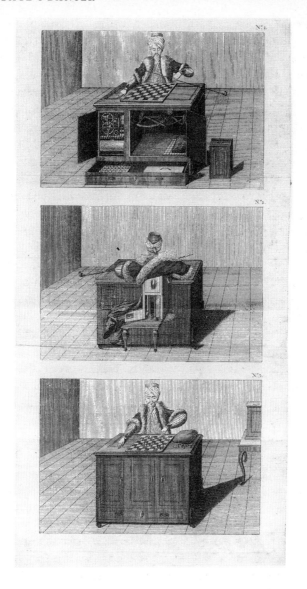

Von Kempelen's Automatic Turk, with the inner gears visible (no.1) and in its playing position (no.3). (*Inanimate Reason*, London, 1784)

Innumerable pamphlets followed the Turk's progress across
Europe to London in the mid-1780s, where, having already been
used in Vienna to bemuse aristocratic visitors to the Imperial court,
and in Paris to contest the mastery of the chess wizards at the Café
de la Régence, it now rivalled the shows of Merlin and Maillardet in
Mayfair. The Turk's arrival in western Europe coincided precisely
with that of another Viennese guru, Franz Mesmer – while von
Kempelen had the ability to build an automaton which displayed
human intelligence, mesmeric séances seemed able to reduce the
most rational humans to the condition of automata. Indeed, von
Kempelen seems first to have built the Turk to distract attention
from Viennese interest in the phenomenon of animal magnetism.
Enlightened philosophers drew the appropriately arrogant moral:
'these days', claimed a gossipy German periodical in 1783, 'physics,
chemistry and mechanics have produced more miracles than those
believed through fanaticism and superstition in the ages of ignorance and barbarism'. Others simply reckoned that von Kempelen
must be dealing with the devil. Like Mesmer, however, the Turk
was also the target of committed exposés. 'The machine cannot
produce such a multitude of different movements, whose direction
couldn't be foreseen in advance, without being subject to the continual influence of an intelligent being.' Some London commentators
immediately alleged the automaton contained a child, or a dwarf,
inside the box, without ever quite managing to explain where the
diminutive prodigy lay hidden, nor did the hostile stories which
appeared in such profusion yet damage the enthusiasm of the automaton's public.

On von Kempelen's death in 1804, the Turk was soon bought by
a brilliant Viennese musical engineer, Johann Maelzel, court mech-
anician for the Habsburgs and a close ally of one of their favoured
composers, Beethoven. Maelzel swiftly saw the patronage he could
win by trading on von Kempelen's automaton, and the Turk became
a temporary habitué of the new Napoleonic courts in Germany.
E. T. A. Hoffmann, a fellow musician, found the figure of a mechanical
Turk a suitably exotic subject for his pen, and in 1814 sent a Leipzig
musical magazine a story entitled 'The Automata', in which he

'took the opportunity to express myself on everything that is called an automaton', teasingly hinting that the Turk might work by setting up a musical harmony with the mind of its audience. Hoffmann used his story to debate the most up-to-date views of occultist German philosophies of nature, much devoted to the inner rhythms of human mental life, then turned his attention to the equally modish attempts to mechanize musical composition. Meanwhile, Maelzel threw himself into another lucrative mechanical project for regulating musical performance with a device he baptized the 'metronome'. The metronome was, of course, a rather more potent means of mechanizing and standardizing artistic creativity than any mere chess player would ever be: 'an universal standard measure for musical time is thus obtained', chorused the musical journalists, 'and its correctness may be proved at all times by comparison with a stopwatch'. After furious patent suits with rival inventors and complex negotiations with Beethoven, Maelzel established himself as the monopoly distributor of these newfangled musical time-keepers, repurchased the Turk from the Bavarian court and then, in 1818, set off on a marketing and publicity tour of Paris and London.

Maelzel's London show was very carefully staged. In autumn 1818 at Spring Gardens, in a pair of candlelit drawing rooms equipped with sloping benches, he displayed the Turk alongside a fine mechanical trumpeter, and when he soon moved round the corner to a larger chamber in St James's Street he added a moving diorama of the burning of Moscow, 'in which Mr Maelzel has endeavoured to combine the Arts of design, mechanism and music so as to produce by a novel imitation of Nature a perfect facsimile of the real scene', and a set of automatic rope dancers, 'scarcely to be distinguished from those of a living performer', moving 'with the utmost correctness without any apparent Machinery'. Maelzel also helped realize his metronomic dream of a completely automatic orchestra, the Panharmonicon of forty-two mechanized musicians, for which Beethoven had specially composed his ghastly *Battle Symphony*, a composition rather likely to appeal to jingoistic British audiences. One London paper praised such an orchestra which 'displayed none of the *airs* of inflated genius, but readily submitted to being wound up'.

The wind-up Turk, of course, occupied pride of place amid these other wonders. Maelzel faithfully followed von Kempelen's recipe, imitating precisely the ritual opening of the cabinet front then back, moving a candle round the interior and winding up the mechanism at regular intervals. And his metropolitan audience faithfully reproduced their earlier enthusiasm for the show. Ever considerate to this public, Maelzel even announced that the automaton would purposely make bad moves so as deliberately to lose if the company seemed bored with over-lengthy games, while the Turk's opponents were ordered to move as fast as possible to alleviate the tedium. A pamphlet authored by a pseudonymous Oxford graduate alleged the whole trick relied on a hidden piece of wire or catgut: 'It seems to be a thing absolutely impossible', the Oxonian alleged, 'that any piece of mechanism should be invented which possessing perfect mechanical motion should appear to exert the intelligence of a reasoning agent'. Unmoved by this futile revelation, during the summers of 1819 and 1820, when the London season ended, Maelzel took his show to the provinces and to Scotland, with apparently equal success. Back in St James's at the end of 1820, however, his nemesis was ready at last.

Robert Willis was an ingenious young Londoner, heir to a distinguished medical family – his father famously attended George III during the monarch's madness. In later life Willis himself became, like Babbage, a pre-eminent Cambridge mathematician, then distinguished professor of applied mechanics and untiring surveyor of ecclesiastical architecture. He also produced one of the best mechanical analyses of the principles of speaking machines, including those von Kempelen had tried to build. During 1819, still a teenager, he patented a new mechanism for harp pedals and toured all the major instrument shops, including high-class machinists such as Holtzappfel and Bramah, and John Newman's works, supplier of equipment to many of the best scientific lecturers in the city. Early in the year he noticed a telling advertisement placed in the London papers by one 'Monsieur Novoski' of Knightsbridge post office, who offered to

MAELZEL'S EXHIBITION,

No. 29, St. James's Street.

The Automaton **Chess Player.**

The Merits of this wonderful Piece of Mechanism is justly appreciated by the Public, it will continue to play several interesting and extraordinary Positions or Ends of Games.

THE AUTOMATON ROPE DANCER.

The Mechanism of which is totally different from, and superior to any thing of the kind ever exhibited in *England*; it will perform several surprising Evolutions: scarcely to be distinguished from those of a living Performer; it sits perfectly free on the Rope, and moves with the utmost ease and correctness, without any apparent Machinery.

THE

AUTOMATON TRUMPETER,

AND THE

Conflagration of Moscow,

In which Mr. M. has endeavoured to combine the Arts of Design, Mechanism, and Music, so as to produce, by a novel Imitation of Nature, a perfect Fac Simile of the real Scene. The View is from an elevated Station on the Fortress of the *Kremlin*, at the Moment when the Inhabitants are evacuating the Capital of the Czars, and the Head of the French Columns commences its Entry. The gradual Progress of the Fire, the hurrying Bustle of the Fugitives, the Eagerness of the Invaders, and the Din of warlike Sounds, will tend to impress the Spectator with a true Idea of a Scene which baffles all Powers of Description.

The Exhibition commences precisely at Two o'Clock, and continues open until Five.

Admission 2s. 6d. *Children* 1s. 6d. *each.*

Parties of Twenty may be accommodated with a PRIVATE EXHIBITION (on previous Application) for Six Guineas, above that number Ten Guineas, when the Chess Player will play whole Games.

Mr. M. begs leave to announce that the Orchestrion, Automaton Trumpeter, Conflagration of Moscow, *and the Patent for the* Metronomes, *are to be disposed of.*

Printed by W. GLINDON, 51, (late 48) Rupert Street, Haymarket.

Maelzel's exhibition of the Automaton Chess Player at St James's Street, 1819. (Robert Willis's Notebook, Cambridge University Library Adv.c.98.14) sell the secret of the chess player for 2,000 guineas. Willis took up the challenge at once. He bought copies of the games the Turk had played and won. He made sure to visit Maelzel's show while it occupied the cramped space of Spring Gardens, 'more favourable to examination as I was enabled at different times to press close up to the figure while it was playing'. Then he smuggled an umbrella into the room so as to measure 'with great accuracy' all the dimensions of the Turk's celebrated cabinet. He went back to St James's for the 1820 season, and by the autumn had completed an obsessively detailed analysis of just how the 'automaton' really worked.

Willis's accurate umbrella and his command of wheelwork made all the difference. The chest, he demonstrated, was much larger than it seemed, giving more than enough room for a fully grown (and doubtless experienced) human chess player to fit inside. 'Instead of referring to little dwarfs, semi-transparent chess boards, magnetism, or supposing the possibility of the exhibitor's guiding the automaton by means of a wire or piece of catgut so small as not to be perceived by the spectators', Willis's *Attempt to Analyze the Automaton Chess Player*, finished in December 1820, proffered the simplest possible scheme of the Turk's hidden intelligence. The noisy gearwheels were there simply so that their sound would conceal any noise made by the concealed player. Even more influentially, he ponderously laid down the law of mechanism's limits:

The movements which spring from it are necessarily limited and uniform, it cannot usurp and exercise the faculties of the human mind, it cannot be made to vary its operations so as to meet the ever-varying circumstances of a game of chess. This is the province of intellect alone.

Despite the care with which he drew the dramatic plates for his pamphlet, there was still some continuing debate about Willis's story: for example, while he reckoned the hidden chess player put his arm inside that of the Turk, more reflective analysts guessed that the machine must use the kind of pentagraph Babbage described in the case of drawing automata. These debates received their fullest publicity in the bestselling *Letters on Natural Magic* (1832) by the Edinburgh optical expert David Brewster. Maelzel himself left London, tried to sell the Turk in Paris, then went to the United States in 1825. Willis's pamphlet was frequently reproduced there by newspapers eager to exploit public interest in the automaton. After Maelzel's show had stunned Richmond and Baltimore, the local writer Edgar Allan Poe lifted Willis's report directly from Brewster, brazenly passed it off as his own brilliant detective job in the *Southern Literary Messenger* (1836), then used it as a precedent for a whole series of rather more original, and certainly better-known, stories about the application of analysis to cunning mysteries, whether purloined letters or concealed bodies. Poe never forgot von Kempelen, nor his deceit, for in the midst of the California gold rush Poe teasingly announced that a New Yorker of that name, doubtless connected with the Slovak engineer, had discovered how to transmute lead into precious metal. If the Turk's progenitor had this shadowy afterlife, its promoter Maelzel never came back to Europe – he died on board ship off Cuba in 1838 – and nor did the Turk, who went up in flames in Philadelphia scarcely fifteen years later.

Moralists found the story of the chess automaton irresistible. The pious drew the obvious implication that if a genius like von Kempelen had not been able to build a rational machine, what must be the skill of the divine Creator who *had* pulled off this trick? Others sang the praises of the Turk's hidden managers, now revealed as William Lewis and Jacques Mouret, who'd played chess so well and under such apparently overwhelming disadvantages in Britain between 1818 and 1820. George Walker of the Westminster Chess Club, writing in *Fraser's Magazine* in 1839, mourned Mouret's 'beautiful emanations of genius' when the Frenchman 'burnt out his brain with brandy' and died in Paris 'reduced to the extremest stage of misery and degradation'. It was a commonplace that such machines belonged at court, whether in the Orient of the Arabian Nights or the grandiose palaces of the Tsars. 'To the half-bred savages of the north', Walker sneered, 'the exhibition could not fail to be striking.' Novels and reviews told how the Turk had conned the powerful and humbled the great: 'even Bonaparte, who made automata of Kings and Princes at his will, was foiled in an encounter

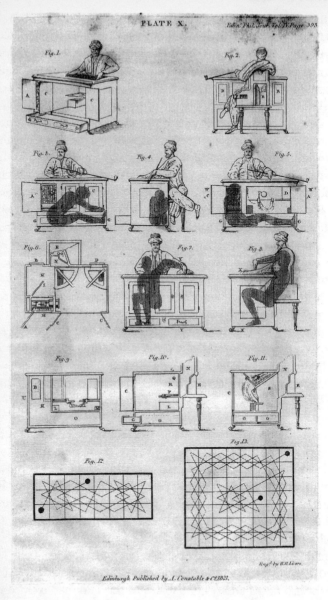

Willis's picture of the concealed chess player. (*Edinburgh Philosophical Journal*, 1821)

with the automaton chess player'. Later in the century, a successful
French play put the Turk on stage in a victorious contest with
Catherine the Great. It was rather the point, so populist writers
explained, that self-styled experts, the politically powerful and the
superstitious mob, could all be deceived by a mechanism effortlessly
unmasked in public prints directed at a new and confidently rational
readership. 'Had the gulled mob reasoned on the matter earlier,'
Walker noted in dangerously republican terms, 'King Automaton
would have been speedily deposed from his high places.'

But the most telling lesson of the Turkish chess player was the
relationship between machine intelligence, technological progress
and the puzzles of concealment. This was a moment when, as the
automaton's admirers never hesitated to remark, 'political economists amuse themselves and the public with the nicely-balanced
powers of man as a propagating and eating *animal* and philosophers
and divines often assure us that he is, in other and higher respects,
but a *machine* of a superior description'. Correspondents from the
London chess clubs predicted that 'a man inside will most assuredly
never again work the charm, but, advanced as science is during the
present generation, a Brunel or a Stephenson could easily and successfully vary the deception'. And in the 1830s one economic journalist, describing the rapid growth and progress of automation in
the Lancashire cotton industry, told the apocryphal story of the
invention of the power loom, half a century earlier, by Edmund
Cartwright. Cartwright had allegedly seen the Turk in London,
trusted its purely mechanical origin, and thus been thoroughly convinced that a weaving machine could scarcely be harder to make
than one which could play chess so well. Rather similar stories
appeared in Poe's favourite source, Brewster's *Natural Magic*, which
was dedicated to Brewster's close friend Walter Scott and entirely
devoted to teaching his fellow citizens the inner secrets on which all
apparently miraculous and surprising mechanical devices really depended. Part of the point was characteristically Presbyterian: gaudy
tricks conned the ignorant into idolatry. Part, however, was economic. In his chapter on automata, Brewster devoted pages to von
Kempelen's chess player, summarized the other notorious stage

shows of the age, then moved straight to Babbage's calculating engines themselves. 'Those mechanical wonders which in one century enriched only the conjurer who used them, contributed in another to augment the wealth of the nation. Those automatic toys,' he concluded, 'which once amused the vulgar, are now employed in extending the power and promoting the civilization of our species.' Apparently theatrical automata really had inspired the Industrial Revolution.

Surrounded by stories which made the intimate link between the plausibility of mechanizing intelligence and the reality of automatic manufacture, it was scarcely surprising that Babbage's new calculating engines, first seriously proposed to the new Astronomical Society in 1820 and first publicized in a pamphlet of summer 1822, should raise in such a lurid way the puzzles of mechanical intelligence which the Turk had just dramatized. It was also rather predictable that his reflections on the intelligence of calculating machines should eventually culminate in a detailed analysis of whether chess could be reduced to a program and an engine whose inner workings would be completely hidden from view behind a gaudy exterior. In his public letter to the President of the Royal Society, Humphry Davy, written barely eighteen months after Willis's pamphlet on the chess automaton, Babbage conceded that his own plans for a machine to 'substitute for one of the lowest operations of human intellect' might 'perhaps be viewed as something more than Utopian, and that the philosophers of Laputa may be called up to dispute my claim to originality'. He was, as usual, absolutely right.

Babbage and his collaborators never found it very easy to teach their audiences just *where* the intelligence of the calculating engines really was. This became even more true when the Analytical Engine was launched in the 1830s. At the end of 1837, Babbage composed a long memoir on its powers, noting that, 'in substituting mechanism for the performance of operations hitherto executed by intellectual labour, the analogy between these acts and the operations of mind almost forced upon me the figurative employment of the same terms'. Phrases like 'the engine *knows*', he confessed, were simply irresistible even though they might be misleading. And the lengthy

Sketch of the Analytical Engine, a joint effort of the early 1840s by Ada Lovelace and Babbage's Piedmontese admirer the mathematician Luigi Menabrea, urged that 'it is desirable to guard against the possibility of exaggerated ideas that might arise as to the powers of the Analytical Engine'. It was a slave which could only follow what it was ordered to do, could originate nothing, was not, in fact 'a thinking being, but simply an automaton which acts according to the laws imposed upon it'.

But there were real dangers in the apparently innocent word 'automaton'. Ada Lovelace, who in 1843 called herself 'the High Priestess of Babbage's Engine', famously explained how it worked by comparing it with the best-known programmable machine of the automatic weaving system, the Jacquard loom, a device whose introduction had almost completely destroyed the jobs of silk weavers in London's East End. 'We may say most aptly', she noted in the *Sketch*, 'that the Analytical Engine weaves algebraical patterns just as the Jacquard loom weaves flowers and leaves.' Lovelace never raised the problem of the substitution of weavers' intelligence by a series of automatic program cards nor the consequent sufferings of London's skilled unemployed. Instead, she directed her polite readers to the Adelaide Gallery, where they would see the loom at work. But such galleries were scarcely likely to resolve the problem of exactly whether the engines could think. In fact, just as in Babbage's reception rooms the Silver Dancer and the Difference Engine stood next door to each other, so at the Adelaide Gallery the Jacquard loom 'in daily operation' stood in the next room to a splendidly automatic Chinese juggler. It was exactly in such places that the distinction between entertaining automata and rational engines was all too easily effaced, and it was there, too, that all the puzzles of mechanized intelligence were graphically put on show.

There is a tempting contemporary resonance to these stories of dancers, Turks, chess and calculating engines. The currently canonical way of telling whether a machine is intelligent explicitly involves deceiving an audience in a manner all too reminiscent of Maelzel's shows at Spring Gardens. The so-called Turing test requires the construction of a hidden device which can produce outputs capable of convincing human judges that it is one of them. Alan Turing, brilliant Cambridge-trained mathematician and veteran of the secret wartime campaign to crack the German Enigma code, was a keen reader of Babbage and Lovelace and much concerned with the problems of automating chess. 'One can produce paper machines for playing chess,' Turing wrote in 1948; 'playing against such a machine gives a definite feeling that one is pitting one's wits against something alive.' In 1950, after a debate at Manchester University on the possibility of making intelligent machines, he wrote a paper proposing what he called an 'imitation game', in which a man and a machine would both feed typed answers to a judge concealed in a different room. Turing's own life was soon to be destroyed by homophobic persecution, so it is intriguing that his first version of the imitation game involved judging which of two invisible respondents was a man, which a woman. In the later test of human and machine, if the judge could not tell which was the man, then the computer would pass the intelligence test. Turing's 1950 paper explicitly discussed Lovelace's ideas about whether the Analytical Engine could be truly intelligent, while in an earlier version he proposed, like Babbage, programming random elements into the computer so as to increase its capacity for innovative intelligence. Turing optimistically predicted that by the century's end computers would have developed so much that they would win a five-minute game at least three times in ten, and public computer competitions, funded by a manufacturer of portable disco dance floors and directed by a behavioural psychologist, are run nowadays in California on exactly the lines Turing set out.

The Turing test is about concealment and detection. Its appeal hinges on the place where intelligence is to be found inside a space to which access is forbidden. In a celebrated paper published in 1980, the philosopher John Searle argued against the conventional interpretation of the Turing test. He proposed a device of which von Kempelen would have been proud, a 'Chinese room' occupied by a human being completely devoid of intelligence about Chinese but supplied with a set of symbols and rules which would allow response to inquiries from outside the room. Searle envisaged that such a system might pass the Turing test by being indistinguishable from native Chinese speakers to anyone outside. But neither necessary nor sufficient conditions could be given for attributing intelligence to such a system. In response to the suggestion that while the room's occupant might not possess intelligence about Chinese, yet the entire system might be said to do so, Searle countered that the enterprise of artificial intelligence 'must be able to distinguish the principles on which the mind works from those on which non-mental systems work', and to judge that a system is intelligent just because of its inputs and outputs would force us to attribute intelligence to a wide range of non-mental systems. The parable of the Chinese room dramatizes the spatial mode of such debates, by insisting on a definition of the place where intelligence might be said to reside.

Here the geography of intelligence is not simply a matter of mundane showmanship, but also relies on the exoticism of distance and the esotericism of concealment. In many Western myths of mechanical intelligence, with Chinese or Japanese, Turks or Nazis as their protagonists, aliens *are* automata, mindless subjects of tyranny; they *build* automata, because they possess fiendish cunning; and they *conceal* what they have done, because they desire to master us. In the world of Babbage's Dancer, mechanical imitation seemed most at home in oriental climes of which rather little was supposed to be known and to which almost anything might be credited. In Turing's world, as his biographer Andrew Hodges points out, intelligence meant secretive messages passed within a guarded coterie and the cryptanalysis of enemy codes as least as much as computers' capacity to imitate human beings. At the end of Gibson and Sterling's fantasy, the cunning orientals would inherit the earth by mastering machine programming and British engines fail through dark conspiracies. There is, perhaps, a longterm political and aesthetic relationship between intelligent automata, orientalism and the covert. The tale of the automata and their impresarios confirms that the places whence machines come, where machines are put on show and the places within machine systems where intelligence is supposed to reside raised, and still raise, delicate political and philo-sophical issues.

Most of these issues hinge on the problem of work and its visibil-ity. Babbage never reconciled himself to the workforce on which he relied; von Kempelen exploited preconceptions about the role of skill; Turing explicitly ruled out any computational task which required the use of a body. Much is made of the collaborative work required from human beings to make their machines look expert and intelli-gent in a recent book by the sociologist Harry Collins, *Artificial Experts* (1990). 'One of the reasons we tend to think a calculator can do arithmetic', Collins suggests, 'is the natural way we help it out and rectify its deficiencies without noticing. All the abilities we bring to the calculation – everything that surrounds what the calculator does itself – are so widespread and familiar that they have disappeared for us.' This is the point of the tale of the Dancer and the Difference Engine. The intelligence attributed to machines hinges on the *cultural invisibility* of the human skills which accompany them. In Babbage's devices, the skills that surrounded automatic mechaniz-ation were systematically rendered invisible. Then and only then might any machines seem intelligent. The moral about the politics and geography of serious trickery is certainly worth remembering. If such machines look intelligent because we do not concentrate on where their work is done, then we need to think harder about the work which produces values and who performs it.

NOTE: The following titles are suggested for further reading: Richard Altick, *The Shows of London* (1978); C. M. Carroll, *The Great Chess Automaton* (1975); Alfred Chapuis and Edmund Droz, *Automata* (1958); H. M. Collins, *Artificial Experts: Social Knowledge and Intelligent Machines* (1990); Anne French, *John Joseph Merlin: The Ingenious Mechanick* (1990); William Gibson and Bruce Sterling, *The Difference Engine* (1988); Andrew Hodges, *Alan Turing: The Enigma* (1983); Anthony Hyman, *Charles Babbage: Pioneer of the Computer* (1982).

The Garden of Edison:
Invention and the American Imagination

PORTIA DADLEY

This jungly garden of old Edison's is a devilish place.
John Updike, *Rabbit at Rest*, 1990

On 1 June 1869 Thomas Edison received the first of his 1,093 patents. US Patent No. 90,646 covered an electric vote recorder, a telegraphic device designed to speed up the voting procedures of Congress. Edison wanted to equip the representatives (currently balloted by lengthy roll-calls) with a simple piece of apparatus which included two buttons, one for a 'yes' vote and one for a 'no' vote. The results of this electromagnetic ballot would be instantly recorded on a frame by the Speaker's desk, two separate dials registering the cumulative totals of 'yes' and 'no' votes. But when he demonstrated his expeditious machine to a Congressional Committee in Washington, the twenty-two-year-old inventor received a cool response from the chairman:

Young man, that won't do at all! That is just what we do *not* want. Your invention would destroy the only hope the minority have of influencing legislation. It would deliver them, bound hand and foot, to the majority. The present system gives them a weapon which is invaluable, and as the ruling majority always knows that it may some day become a minority, they will be as much averse to any change as their opponent.

Reporting these remarks in an interview for *Harper's New Monthly Magazine* in February 1890, Edison described the impact of the chairman's no-vote: 'I . . . was about as much crushed as it was possible to be at my age. The electric vote recorder got no further than the Patent Office.'

This inauspicious début proves the truth of Alfred North White-head's observation in *Science and the Modern World* (1926) that one of the challenges facing the inventor is to bridge 'the gap between the scientific ideas and the ultimate product'. In this respect the vote recorder functions as a symbolic prototype of Edison's career: it is a faulty arrangement of scientific and social components that will be corrected, refined and reassembled by the mature inventor. One of the adjustments to be made is a linguistic one. The vote recorder's binary code must, somehow, connect up with the elaborate negotia-tions of the filibustering congressmen: Edison must adapt his me-chanical formula to the shifts and nuances of democratic discourse.

The connection Edison needed to forge between language and technology was one with a long pedigree in American history. Article 1, Section 8 of the United States Constitution states:

The Congress shall have Power . . . To promote the Progress of Science and useful Arts, by securing for limited Times to Authors and Inventors the exclusive Right to their respective Writings and Discoveries.

This law of patent reads awkwardly today: it was not until 1829 that the Harvard physician Jacob Bigelow chose the 'general name of Technology' to describe 'the principles, processes, and nomencla-tures of the more conspicuous arts, particularly those which may be considered useful'. Like the US Constitution, this definition (given in Bigelow's book *Elements of Technology*) creates an alliance be-tween art and science. For if technologists built machines that were socially 'useful', they also contributed to a fairy-tale vision of progress. 'The labor of a hundred artificers is now performed by the operations of a single machine,' boasted Bigelow. 'We traverse the ocean in security, because the arts have furnished us a more unfail-ing guide than the stars. We accomplish what the ancients only dreamt of in their fables; we ascend above the clouds, and penetrate into the abysses of the ocean.'

Technology helped to make America come true. As the wilderness fell to the axe, plough and engine, the nation's political contours became clear. In his Sixth Annual Message (December 1806) Thomas Jefferson supported the development of new transportation technologies, arguing that the 'channels of communication' created by a road and canal network would cement the union and make 'lines of separation disappear'. Edison's achievements need to be set against this republican landscape; his inventions were embedded in the democratic tradition he celebrated in the preface to a new edition of *The Works of Thomas Paine* (1934). 'The man had a sort of universal genius,' wrote Edison of the innovator who drew up the revolutionary *The Rights of Man* (1791–2) as well as plans for an iron bridge, a hollow candle and a draught burner. Paine's comprehensive talents were shared by his admirer; Edison pioneered technologies that aided and entertained every segment of the population.

America's collective 'yes' was the secret of the invention that made Edison a household name. On 24 December 1877 Edison applied for a patent, covering a method of

arranging a plate, diaphragm, or other flexible body capable of being vibrated by the human voice or other sounds, in conjunction with a material capable of registering the movements of such vibrating body by embossing or indenting or altering such material, in such a manner that such register marks will be sufficient to cause a second vibrating plate or body to be set in motion by them and thus reproduce the motions of the first vibrating body.

This complex arrangement of vibrating bodies was ratified with Patent No. 200,251 in February 1878; but Edison was able to syndicate its popular usage almost immediately. On 7 December 1877 Edison took his new machine to the offices of the *Scientific American* in New York. On 22 December an editorial appeared in the magazine describing the upshot of this visit:

Mr Thomas A. Edison recently came into this office, placed a little machine on our desk, turned a crank, and the machine enquired as to our health, asked how we liked the phonograph, informed us that *it* was very well, and bid us a cordial good night.

This simple greeting (a marked contrast to the recondite specifications of Edison's patent application) was the opening gambit in a dialogue that the inventor would conduct with the American public for the next fifty years. The phonograph supplied the social deficiencies of Edison's doomed vote recorder: here was a device that permitted the consumer to leave his or her mark on the products of technology. An assortment of linguistic experiments was performed on Edison's new invention. There is the American professor who used the phonograph 'to secure good records of cat language' ('It is not difficult to secure the record of an angry cat's voice, for all you have to do is to hold the animal near the mouth of the phonograph and give its tail a twist'); the disembodied voice of a London coster-monger that 'proceeded apparently from the very midst of the vegetables' on his barrow, advertising 'termarters' at 'tuppence a pahnd' and 'green peas fippence the 'alf peck'; and the 'praying machine' of Lhasa, a phonograph imported into Tibet in 1897 and 'kept busy with [the Dalai Lama's prayers] and other utterances holy to the Buddhists'.

These peculiar experiments point to a fundamental paradox of late nineteenth-century technology. The phonograph, like many of Edison's inventions, relieved the drudgery of everyday life and trans-formed routine into something new and strange: *Frank Leslie's Illustrated Newspaper* praised the 'nineteenth-century miracle' that would 'turn all the old grooves of the world topsy-turvy and estab-lish an order never dreamed of even in the vivid imaginings of the Queen Scheherezade'. But this fantasy was offset by what Daniel Boorstin, in *The Americans: The Democratic Experience* (1973), has described as 'the technology of repeatable experience'. Not only did the phonograph replay its recorded messages *ad infinitum* (in an interview with the *Washington Post* Edison pictured his invention as a 'tongueless, toothless instrument' which 'will repeat again and again to a generation that will never know you, every idle thought, every fond fancy, every vain word that you choose to whisper against [its] thin iron diaphragm'); but it also became a mass-pro-duced object. In 1878, less than a year after Edison carried his model into the offices of the *Scientific American*, the Edison Speaking Phonograph Company was set up to manufacture and market the new device.

Edison camouflaged the impersonal aspects of his work by casting the phonograph in his own image. Talking to the *New York Graphic* in April 1878 he announced: 'I've made some machines, but this is my baby, and I expect it to grow up to be a big feller, and support me in my old age.' Edison made sure of this technological coming of age by nourishing his filial metaphor (a commonplace now among R & D workers) at every available opportunity. In the interview with the *Graphic* he gleefully fleshed out the scientific skeleton of the phonograph by suggesting that he might 'substitute some sort of membrane for this ferrotype-tympanum, and put some sort of a voice chamber over the mouthpiece about the size of the human mouth, with teeth and perhaps tongue. This will give the resonance that is lacking in the machine.' That June, Edison outlined some chores for his vociferous offspring in an article for the *North American Review*. A ten-point list of suggested uses for the phono-graph included clocks that 'will tell you the hour of the day, call you to lunch, send your lover home at ten, etc'. The ghost of this obedient child haunted Edison's own household. 'Eleven o'clock, one hour more,' a phonographic clock was said to have called out to an unsuspecting guest; and, one hour later, 'Twelve o'clock, prepare to die.'

   '*And his singular whisper it grew the very echo of my own.*' Like the beguiling accents of the double in Edgar Allan Poe's story 'William Wilson' (1839), the phonograph cast its spell by reproducing, with absolute fidelity, the familiar sound of the human voice. So, the more Edison 'domesticated' his invention, the stranger it became. Resonances of the uncanny are captured on a phonogram that Edison sent to Colonel George Gouraud, his London agent, in June 1888, shortly after the birth of his fourth child, Madeleine. In send-ing the phonogram (the world's first) Edison not only provided Gouraud with a sample of the improvements he was making to the technology of the phonograph, but clarified his paternal priorities:

Friend Gouraud

   Ahem! This is my first mailing phonogram . . . I send you by Mr Hamilton a new phonograph, the first one of the new model which has just left my hands.

   It has been put together very hurriedly, and is not finished, as you will see. I have sent you a quantity of experimental phonogram blanks, so that

Colonel Gouraud and his family listening to Edison's messages. (From W.
K. L. Dickson and Antonia Dickson, *The Life and Inventions of Thomas Alva
Edison*, London, Chatto & Windus, 1894)

you can talk back to me . . .

Mrs Edison and the baby are doing well. The baby's articulation is quite loud, but a trifle indistinct; it can be improved, but is not bad for a first experiment.

With kind regards,
Yours, EDISON

This communiqué fulfils the promise of the primogenital metaphor delivered to the *Graphic* ten years earlier. By mixing up the new arrivals in the Edison household, the proud father personalizes the invention 'which has just left my hands'; but, by the same token, he erases the identity of his wife and child. Like the 'experimental phonogram blanks' Mina Edison and her unnamed, ungendered child are featureless bodies, filiated to the inventor only by impress of his linguistic ingenuity. Edison's two elder children, Marion and Thomas Jr, were components in the same process. Under their nicknames 'Dot' and 'Dash', they were instantly recognizable as members of America's symbolic family of science.

Edison's neglect of marital and parental duties is frequently noted by his biographers, but perhaps it is truer to say that the inventor's private and public lives were inextricably linked. Edison was what he invented, and invented what he was. He twinned his brand-new daughter and rejuvenated phonograph in an effort to generate publicity for his latest technical project. Gouraud's phonogram became one of the attractions of the Crystal Palace Exhibition: it was even played for Prime Minister Gladstone at a special preview. Edison's generic wordplay was directed at potential customers: he stamped a standard mass-produced object with his own highly individualized image. The role of branding in product development is symbolically embodied in the procedure that Edison, who was nearly deaf, adopted for testing the phonograph. When the sounds were faint, the inventor would pick up the vibrations by biting into the wood of the machine, thus simultaneously absorbing himself in his work and attuning himself to the exigencies of mass consumption.

If Edison's improvisations encouraged a naïve belief in the artless character of his inventions, then so too did the tales of his boyhood that began to appear in the national press at the end of the 1870s. In these tales the young inventor featured as a Huck Finn-like character, his rough-and-ready exploits making a gripping narrative of scientific development. In one story 'Al' Edison perpetrated a babyish balloon hoax by feeding his playmate Michael Oates with Seidlitz powders (a kind of laxative), in the expectation that his gas-filled subject would fart his way into orbit; in another, the embryonic inventor rigged up a primitive telegraph, and tried to generate static electricity for the line by rubbing two tom cats together. These exploits – which soon acquired legendary status – gave cultural momentum to the scientific discoveries and technical innovations that had reshaped the patterns of nineteenth-century American life. Like *The Adventures of Huckleberry Finn* (1884), the adventures of Al Edison depicted a retrospective idyll. While the mature inventor was propelling America into the future with his technically advanced creations, his juvenile *alter ego* extemporized a reassuringly unsophisticated interpretation of the nation's recent past.

Certainly there were some objective facts in these dramatic accounts ('Tom Alva never had any boyhood days,' said his father. 'His early amusements were steam engines and mechanical forces.'), but Edison's talent for engineering raw biographical material into a compelling narrative also illustrates his definition of the inventor as 'the specialist in the high pressure stimulation of the public imagination'. In the most famous episode of his childhood, the fifteen-year-old Edison, operating a concession to sell newspapers and refreshments on the Grand Trunk Railway, gradually filled the baggage compartment where he kept his stock with a collection of jars, bottles, batteries and even a small printing press – used to produce the *Weekly Herald*, a newspaper written, edited and distributed by the dynamic newsboy. A lurch from the train, a toppling jar of phosphorus, and the compartment was ablaze: boy, laboratory and newsroom were thrown off at the next station.

This story, with its combustible mix of science and newsprint, touches off a major theme in Edison's career. Experiment, for the boy inventor, is inseparable from publicity: from his earliest years Edison combined investigation with the messages and dispatches that were reshaping the reading habits of the American public. His prowess in the field of information technology is demonstrated in another incident from 1862. Picking up news of the battle of Shiloh during a stopover at Detroit, Edison wired ahead to the 'station-agents, who were also telegraphers', 'asking them to post notices that when the train arrived I would have newspapers with the details of the great battle'. '[The] device', the inventor recalled, '. . . worked beyond my expectations.' Spying a throng of eager buyers at the first stop, the young entrepreneur made a calculating move:

After one look at that crowd I raised the price from five cents to ten and sold as many papers as the crowd could absorb. At Mount Clemens, the next station, I raised the price from ten cents to fifteen. The advertising worked as well at all the other stations. By the time the train reached Port Huron I advanced the price of the Detroit Free Press for that day to thirty-five cents per copy and everybody took one.

Another version of this anecdote ends at a Port Huron prayer meeting: 'In two minutes . . . the meeting was adjourned, the members came rushing out, bidding each other for copies of the precious paper. If the way coin was produced is any indication, I should say that the deacon hadn't passed the plate before I came along.' Though stretching the bounds of belief, this unholy mêlée provides a ceremonial climax to young Al's technological rite of passage. The body of the Edisonian faithful pay homage (and hard currency) to the messianic figure, who, as Alan Trachtenberg has observed in *The Incorporation of America* (1982), 'seemed to hold together the old and new, the world of the tinkerer and the world of modern industry; the age of steam . . . and the coming age of electricity'.

Edison's supernatural powers were celebrated in his popular incarnation as 'The Wizard of Menlo Park'. The title, which began to appear in the national press shortly after the invention of the phonograph, seems oddly unsuited to the carefree figure of legend, but conjures up the fascination that Edison held for the American public. The wizard, a symbolic figure of science, is powerless without his spells. Captivating the nation with his phonographic performances and tales of adventure, Edison gave a doubly convincing display of semantic sorcery.

Those verbal chimeras were turned into realities at Menlo Park, his laboratory from 1876 to 1887. Contemporary pictures and photographs reveal an unimposing collection of wooden huts sealed off from the surrounding woods by a tidy picket fence. In these rustic scenes Menlo's industrial enterprise is literally out of sight, the activity of its workers contained within endless whitewashed walls. In the seasonal scene opposite, only a row of spindly telegraph posts, barely distinguishable from the nearby trees, spoils the pastoral and unpeopled view.

Like the lengths of track and wire running through young Al's thrilling adventures, this half-hidden line of communication links Edison's rural retreat to the industrial world: it anchors an idealized scene of invention to an efficient network of technological production. 'Mr Edison lives in a world of his own,' revealed Henry Ford in *My Friend Mr Edison* (1930), 'but he knows exactly what is going on in the rest of the world.' So the telegraph connected Menlo with the commercial markets of New York less than thirty miles away and would update potential consumers on Edison's innovative activities. Road and rail links gave access to the industrial resources of the neighbouring towns of Newark and Patterson, and brought crowds of admiring visitors to the 'village of science'. 'When the public tracks me out here I shall simply have to take to the woods,' quipped the inventor when he first came to Menlo; but, like Henry Thoreau self-consciously building his hut within range of his fellow 'townsmen' in *Walden* (1854), Edison took care to signpost his idyllic hideaway. Menlo's vaunted seclusion – celebrated in titles such as 'the tabernacle' and 'the Vatican of science' – was an invention of worldly American journalists. The laboratory was so well publicized that it became a sort of holiday resort, with the Pennsylvania Railroad organizing excursions for thousands of techno-tourists.

One of the Edison sightseers, travelling to Menlo on a dark December night in 1882, has left a superb account of the laboratory's dazzling *mise-en-scène*:

Menlo Park in the winter of 1879. (From W. K. L. Dickson and Antonia Dickson, *The Life and Inventions of Thomas Alva Edison*, London, Chatto & Windus, 1894)

I cannot tell how long we had been rolling along, for, lulled by the movement
of the carriage and buried in my warm furs, I was quietly dozing, when a
formidable, 'Hip, hip, hurrah!' made us all jump, my travelling companions,
the coachman, the horses, and I. As quick as thought the whole country
was suddenly illuminated. Under the trees, on the trees, among the bushes,
along the garden walks, lights flashed forth triumphantly.

So Sarah Bernhardt made her grand entrance into Menlo Park, as
she recounted in *My Double Life* (1903). The run-of-the-mill visitor
would be treated to more modest entertainments (a popular item
was a phonographic duet recorded over a background of crowd
noise, the singers having to contend with shouts of 'Oh! shut up!',
'Go away, if you can't sing any better' and 'Help! Police! Murder!'),
but Bernhardt's Gala Night serves – literally – to illuminate the
theatrical ambitions of '*le grand* Edison'. The inventor-cum-impresa-
rio transformed his laboratory into a staging area for science: Menlo
was a three-dimensional advertising space displaying the latest
Edisonian wares.

This PR exercise held certain perils for the unsuspecting visitor,
as Bernhardt discovered when she went inside the lab. Overwhelmed by 'the deafening sound of machinery' and 'the dazzling
rapidity of the changes of light' the actress loses her sense of direction: 'All that together made my head whirl, and forgetting where I
was, I leaned for support on the slight balustrade which separated
me from the abyss beneath.' Bewitched, as a spectator, by the shining trees and bushes in Edison's garden of light, when she goes
behind the scenes of invention, Bernhardt is thrown out of gear by
her threatening surroundings. Like Georgiana, the heroine of Nathaniel Hawthorne's short story 'The Birth-Mark' (1843), who is dismissed by her husband as a 'prying woman' when she wanders
into his laboratory, Bernhardt oversteps the limits of her sex: Edison
remembered his visitor as a 'terrific "rubberneck"' who 'jumped all
over the machinery' and 'wanted to know everything'. The actress
stands at the threshold of an alternative Menlo, off-limits to the
casual visitor, where the creative work of invention is pursued.

The very secrecy of this inner sanctum makes it difficult to reconstruct Edison's working practices. While the Wizard of Menlo Park

happily conjured a make-believe world of invention for his adoring public, he jealously protected the formulae and procedures that lay behind his spellbinding act. But occasionally the mask would slip, as Sir William Preece, electrical consultant to the British Post Office, discovered in 1878. (Preece's dealings with Edison are described in biographies of both men, notably Ronald W. Clark's *Edison: The Man Who Made The Future* (1977) and E. C. Baker's *Sir William Preece, F. R. S., Victorian Engineer Extraordinary* (1976).) When he visited Menlo in 1877 Preece had received a typically bizarre welcome, noting in his diary for 17 May: 'Another blazing day which I spent at a place called Menlo Park with Edison – an ingenious electrician – experimenting and examining apparatus. He gave me for dinner *Raw ham*! tea and iced water.' This choice entrée was to be followed by rather more unpalatable dishes. In May 1878, when David Hughes, an acquaintance of Preece, announced the invention of a microphone that featured some of the characteristics of Edison's loud-speaking telephone receiver, the American inventor sent a letter to the *New York Tribune*, accusing Preece of betrayal:

I freely showed [Preece] the experiment I was then making, including the principle of the carbon telephone and the variability of conducting power of many substances under pressure. I made him my agent for the presentation of this telephone, and subsequently of the phonograph, in England, and kept him informed, by copies of publications and by private letters, of my leading experiments, as he always manifested a great desire to be the means of presenting my discoveries to the British public. I therefore regard the conduct of Mr Preece in this matter as not merely a violation of my rights as an inventor, but a gross infringement of the confidence obtained under the guise of friendship.

The language of this letter is the antithesis of the 'Aw shucks' tone adopted for the phonographic soundbites and childhood anecdotes. Where Edison lured the American public with the siren voice of invention, here he defends his scientific territory in writing, constructing a closely argued and syntactically intricate case against the offending Preece. Even the evidence Edison uses to verify his claims is text-based: 'copies of publications and . . . private letters' kept Preece informed of the inventor's 'leading experiments'.

Writing sets up a boundary between the professional and popular aspects of invention, supplying the codes that divide initiates from technological outsiders. Eighteenth-century renaissance man Benjamin Franklin told readers of his *Autobiography* (1818) that he was content to let his scientific 'papers shift for themselves' in order to spend time 'making new Experiments'. But by the late nineteenth century the theoretical and practical aspects of invention had become inextricably linked, with periodicals such as the *Scientific American* publishing ideas and information to be reworked by the technological imagination.

'Books lay promiscuously about,' reported Menlo worker Francis Jehl to Henry Ford in *My Friend Mr Edison*. At West Orange, where Edison built a new laboratory in 1888, there was a library containing over ten thousand books and periodicals. According to W. H. Meadowcroft, author of *The Boy's Life of Thomas Edison* (1911), this collection included:

magazines relating to electricity, chemistry, engineering, mechanics, building, cement, building materials, drugs, water and gas power, automobiles, railroads, aeronautics, philosophy, hygiene, physics, telegraphy, mining, metallurgy, metals, music, and other subjects; also theatrical weeklies, as well as the proceedings and transactions of various learned technical societies.

Compiled with bureaucratic zeal (Meadowcroft was Edison's private secretary), this list is calculated to appeal to a juvenile audience, for whom the enumeration of new technologies signified romance and adventure. Witness the success of Victor Appleton's *Tom Swift* books (c. 1910), in which the eponymous hero embarks on one mechanical exploit after another. *Tom Swift and His Motor Boat* and *Tom Swift and His Air Glider* are two of the titles in the series; other inventions mastered by the boy scientist include the airship, the electric rifle and the wizard camera.

Edison (whose own childhood provides a blueprint for Swift's derring-do) also turned literature into an arena where legendary feats were performed. In an interview reprinted in *The Diary and Sundry Observations of Thomas Alva Edison* (1948), he characterized his work as a Herculean struggle with words:

When I want to discover something, I begin by reading up everything that
has been done along that line in the past – that's what all these books in the
library are for. I see what has been accomplished at great labor and expense
in the past. I gather the data of many experiments as a starting point, and
then I make thousands more.

Edison's lengthy researches are celebrated in another experimental epic, *U.S.A.* (1938). John Dos Passos produces a biographical
snapshot of the inventor that plays on variations of the sentence,
'Whenever he read about anything he tried it out.' Whenever he
did try anything out, Edison would set down the results in his lab
notebooks. Figures, notes, sketches, even physical substances, went
down on paper: when he was making experiments for the filament
of his electric light bulb in the late 1870s, Edison attached a fragment of every material he tested, including strands of platinum,
iridium, boron, threads treated with plumbago and coal tar, different
grades of cardboard and drawing paper, wood splints, cornstalks,
over a hundred varieties of bamboo and a red hair from the beard
of J. U. Mackenzie, the station master at Mount Clemens, who had
dropped by one day to catch up with the fortunes of the teenager
he had instructed in the principles of telegraphy. By the time of
Edison's death in 1931 the notebook ran to 3,400 separate volumes;
the miscellaneous activities of the lab were preserved for posterity
in the chronicle that the press described as the 'Edisonian Book of
Genesis'.

Edison was keen to develop his mythic powers of creation. He
told George Parsons Lathrop (son-in-law of Nathaniel Hawthorne)
that the notebook was his 'novel', and later collaborated with Lathrop on *Progress*, a story which explores the fantastic possibility of
space travel (a vehicle is invented which will take man to Mars)
and genetic engineering (a research station on the upper Amazon
breeds apes capable of conversation). *Progress* never reached publication, but the ideas Edison jotted down in his manuscript notes were
reflected in contemporary fiction: interplanetary adventure was the
subject of Garrett P. Serviss's *Edison's Conquest of Mars* (1898), and

in *L'Eve Future* (1886, translated by Marilyn Gaddis Rose, 1981)
Count Villiers de l'Isle-Adam turned Edison into a technological
Adam, responsible for the production of a beautiful female automaton named Hadaly whose lungs are 'two gold phonographs' programmed with 'seven hours of speech'.

The exploits of the make-believe Edison offer a telling insight into
the uncanny qualities ascribed to the flesh-and-blood inventor. 'In
Europe and America popular imagination has built a legend around
this great American citizen,' wrote Villiers in the foreword to his
novel. '. . . So doesn't it follow that the "Personage" of this legend
belongs to popular literature even during the lifetime of the man
inspiring it?' Edison himself gave this question an intriguing twist
when he began experimenting with the new technology of film. In
one 1897 short we see him at work in his laboratory, barking orders
to his assistants, only to be disrupted by the thunder of battle as
uniformed horsemen (portraying the British and Boers in South
Africa) go charging by outside the window. (A more sinister aspect
of Edison's character was revealed in 1910, when he produced the
first Frankenstein movie, starring Charles Ogle as the monster.)

These screen ventures make an apt coda to a career that was
built on the manipulation of images. From his youth Edison had
entranced the American public with idealized versions of his life
and work; film provided a fresh medium for mobilizing the scraps of
narrative he had always used to promote his inventions. One of his
most valuable contributions to the motion-picture industry was the
vitascope (1896), a 'screen machine' which could process long strips
of film through the inclusion of a small reel called the Latham loop.
Early films, only seconds long, had done no more than present the
wonder of preserved movement; hence the attraction of the first
copyrighted movie *Record of a Sneeze* (1894), a self-explanatory
vehicle starring an Edison lab worker named Fred Ott. But with
the vitascope film makers could produce features lasting several
minutes: a time span which permitted the construction of action
sequences and, more importantly, an accompanying plot.

As America fell under the spell of the movies, Edison's own powers
began to wane. 'It is clear that my usefulness is gone,' he confessed

to Henry Villard, President of Edison General Electric, in February 1890, as rumours spread of an impending merger with the Thompson-Houston company. When the takeover was finally completed in February 1892, the *New York World* ran the headline: 'MR EDISON FROZEN OUT/HE WAS NOT PRACTICAL ENOUGH FOR THE WAYS OF WALL STREET.' Sharp practice also came into play when the vitascope appeared on the market. The machine's revolutionary technology was pioneered by an amateur inventor named Thomas Armat. Edison's Kinetoscope Company bought up the original patent and then, as distributing agent Norman C. Raff later explained to a disgruntled Armat, linked the new acquisition to 'Mr Edison's great name' 'in order to secure the largest profit in the shortest time'.

Like the figures that moved across the nation's cinema screens, Edison, at the turn of the century, was reduced to a shadow of himself. Consumed by his own legend, his 'great name' was invoked only to add lustre to the output of the new industrial conglomerates. Playacting, of course, had always been integral to Edison's success, but now his entire career turned into a public spectacle. Edison gave the performance of his life on 21 October 1929, when Henry Ford decided to combine the opening of his new museum of industrial history at Dearborn, Michigan, with 'Light's Golden Jubilee', a celebration of the fiftieth anniversary of the invention of the incandescent light. Ford's museum contained numerous reminders of America's past – a replica of Independence Hall, the Logan County Courthouse where Abraham Lincoln practised law, a blacksmith's forge (an antique in the era of the Model T) – but its centrepiece was Edison's Menlo Park laboratory, which had been transported *in toto* from New Jersey. On the evening of the opening day, the eighty-two-year-old Edison, having already enacted an episode from his youth by selling fruit, sandwiches and newspapers from a replica of the Grand Trunk Train to the assembled dignitaries ('I'll take a peach,' said President Hoover), returned to his old haunt and demonstrated how he made the electric light from a piece of carbonized cotton and a glass bulb. The 'invention' was broadcast, via a gigantic radio hook-up, all over America. Listeners who had tuned into history were urged to sit in darkness, and then switch on their lights when the signal was given. As Edison touched his magic wire, Graham McNamee, who was reporting the event for NBC, shouted into his microphone: 'And Edison said: "Let there be light".'

Sliding Scales: Microphotography and the Victorian Obsession with the Minuscule

MARINA BENJAMIN

John Benjamin Dancer is not a name to be reckoned with in the annals of science. Reading the various biographical notices written since his death in 1887, one is struck with a certain sense of pathos; not even the liberal sprinkling of well-meaning hyperbole endemic to biographical memoirs of scientific societies can disguise the salvage exercise. Here was a man who almost discovered ozone, failed to patent a number of ingenious optical and mechanical devices that might have made him a fortune, improved other people's discoveries rather than made his own, an optician who lost his sight and died courting penury. In short, a man whose career was a catalogue of near misses, bad management and consequential blunders. J. B. Dancer never left the byways of science to join the ranks of celebrated originals like Darwin, Brewster, Dalton and Joule, all of whom he knew and corresponded with. But minor-league science possesses an interest of its own; among wrong-footings and false leads, the arcane products of excessive specialization and the occasional whimsical detour, we may gain insight into cultural obsessions and historical impulses whose significance has been lost amidst grander scientific happenings. And J. B. Dancer may take sole credit for one such scientific detour, the microphotograph, which owed its genesis to a romance with the minuscule and its demise to a disenchantment with things minute.

Microphotographs were the strange fruit of a cross-fertilization between two of the nineteenth century's leading technologies, microscopy and photography. Dancer was an adept in both. A third-generation optician and instrument maker, he early spotted the commercial potential of affordable quality achromatic microscopes which he supplied to numerous medical schools and Mechanics Institutes

in the 1840s and 1850s. When Fox Talbot announced his invention
of photogenic drawing in 1839, Dancer immediately embarked on
his own investigations of the 'Black Art' and was soon producing
photosensitive paper for commercial sale. And when, shortly after-
wards, Daguerre's silver-plate process was publicized in England, he
rapidly succeeded in obtaining pictures which became the first
Daguerreotypes to be exhibited in Liverpool, where his business
was based before he moved to Manchester in 1841. Dancer dabbled
in the possibility of combining microscopy with photography from
the start. During a lecture at the Mechanics Institute in Liverpool,
before an audience of 1,500 people, he made a Daguerreotype image
of a flea magnified to six inches in length. These enlarged images of
microscopic objects were known as photomicrographs, as distinct
from the production of microscopic photographic images or microphotographs. Dancer attempted the latter using Daguerre's method
but found that the size of the mercury particles which formed the
picture limited the reduction. It was only with Scott Archer's development of the wet collodion process in 1851 that he was able to
produce successful microphotographs, which by virtue of being reproducible became commercially viable.

Mounted on standard 3 × 1 glass slides, microphotographs look
deceptively like histological preparations, that is, ultra-thin slivers
of living tissue, but when magnified 100 times, the inscrutable tiny
black dot glued in place under the cover glass is revealed to be an
exquisite, fine-grained reproduction of Raphael's Madonna or the
ruins of Tintern Abbey, not a delicate tranche of liver or a cluster
of blood platelets. Depictions of nature, in any guise other than a
photographic landscape, are conspicuous by their absence. Instead,
what Dancer's microphotographs offer, at least to the contemporary
observer, is an index of Victorian cultural values. Their subjects
range from portraits of the great and good – eminent scientists,
European royals, political and military dignitaries, literati and thespi-
ans; celebrated paintings; religious texts, like the Lord's Prayer or
the Sermon on the Mount; extracts from Tennyson, Dickens, Milton,
Byron and Pope; to views from around the world (forerunners of
the tourist snapshot). The idea that microphotographs represent

Victorians looking at Victorians is reinforced by a number of self-congratulatory images of modern buildings such as Joseph Paxton's Crystal Palace, of monuments and tablets commemorating national heroes, and by their celebration of the era's achievements. There are microphotographs of the several bridges criss-crossing the Seine, the bridge at Prague and the suspension bridge at Conway Castle, as well as an image of a conference of engineers gathered at the Menai Straits to plan floating one of the tubes of the Britannia Bridge; one of the eighty-pounder Whitworth gun and another of Lord Rosse's telescope. Images of faraway places where Britons had alighted (and, more often than not, planted Union Jacks), pictures of maps and of banknotes from around the globe further testify to an interest in exploration and empire. It is difficult to imagine a more thorough-going taxonomy of the cares, concerns and ambitions of the era. Yet how the images might have been read by those who consumed them is another matter entirely and one which this essay will seek to fathom.

Dancer produced his first commercial slide in 1853 – a rather austere picture of electrician William Sturgeon's memorial tablet. By 1873 he was advertising nearly 300 microphotographs and by the end of his career the grand total had risen to over 500. Precisely how he manufactured his microscopic marvels remains a trade secret, since he never ventured into print on the subject. It is known that in experimental trials he used the eyes of recently killed oxen as photographic lenses and that he began the process with 4 × 5 inch collodion glass-plate negatives, but beyond that it can only be assumed that his method of reduction bore some similarity to that publicized by George Shadbolt in 1857. At the time Shadbolt was President of the Microscopical Society and editor of the *Photographic Journal*, in whose pages a priority dispute over the invention of micro-photography took place, Dancer winning the day. Perhaps by way of supporting his own claim to an independent invention of micro-photography, Shadbolt provided the journal with a full description of his experimental set-up complete with an engraving, thus reveal-ing his equipment to be essentially a compound microscope in re-verse. The glass slide coated with wet collodion emulsion – in effect a sort of technological *tabula rasa* – would be placed over the stage aperture of the microscope and an image would be projected on to it by means of an objective lens placed where the condensing lens would normally be, the rest of the microscope being used merely to check the focus. The negative, meanwhile, would be interposed between the microscope and the light source, the light reaching it via two condenser lenses. Once up and running, a succession of glass slides could be speedily imprinted.

Almost as soon as Dancer perfected the mechanics of reproduction, he began selling microphotographs as novelty items. At a shilling a slide, and with decent parlour microscopes to be had for a few pounds, microphotographic entertainment was an economic form of rational recreation. When George Shadbolt disclosed his method of reduction in the *Photographic Journal*, he remarked on the popularity of microphotographs in tones of near surprise: 'the demand for a supply of these minute productions has been so great amongst those possessed of a microscope, that they have become a regular article of manufacture'. In fact, the market for microphotographs was sufficiently sizeable to make it profitable for Dancer to sell his slides to a number of retailers of scientific instruments.

It is somewhat surprising to find that so innocuous an invention as microphotography had its detractors, but criticism came from various high-minded members of the photographic community who took it upon themselves to crusade on behalf of photography as high science. Chief among them was Thomas Sutton, editor of a rather sniffy journal called *Photographic Notes*, and inventor of numerous pieces of camera equipment. In his *Dictionary of Photography* (1858), he dismissed microphotography as being of 'little or no practical utility'. The production of minuscule images was, he claimed, 'a process that must strike any reasonable person as somewhat trifling and childish when he considers how many valuable applications of photography remain yet to be worked out'. Dancer had made an adversary of Sutton in 1852 when he had challenged Sutton's assessment of the correct distance that should obtain between two cameras in the taking of stereoscopic photographs. Sutton proposed a minimum distance of eight inches, while Dancer

George Shadbolt's apparatus for manufacturing microphotographs. The light source [a] is a small camphine lamp that illuminates the negative; [b] is a bull's-eye lens, arranged so that the refracted light rays just fill the whole of the double-concave lens [c], which is placed so as to refract light rays in a parallel direction on to the negative [d]. The negative is placed 2 to 4 feet from the lower object-glass of the microscope within a screen of wood or card that cuts off extraneous light. The objective [e] intended to produce the picture is located on the substage of the microscope. A wet collodion-coated glass slip is placed between [e] and another objective [f], which is used to focus. (From *The Photographic Journal*, vol. 4, 1857. Reproduced by permission of the Royal Photographic Society, Bath) argued that the correct distance was the interocular one of roughly two and a half inches. He went on to construct a twin-lens stereoscopic camera in order to settle the issue. Sutton probably relished the advent of microphotography as an excuse to vent spleen. Shadbolt too, his nose out of joint over the priority dispute, demonstrated in an 1859 issue of his journal that he was not above sour grapes: 'microphotographs can never be more than amusing curiosities,' he declared. The following year the *Photographic Journal* carried a review of James Nicholls's *Microscopic Photography; its Art and Mystery* (1860), that berated the author for not considering that branch of microscopic photography (namely photomicrography) that 'would have rescued the art from the "toy"-sarcasm'.

But microphotography also won fans. Sir David Brewster, who in the 1850s was Professor of Physics at St Andrews, saw streams of possibilities emanating from Dancer's invention. In an article on the micrometer for the eighth edition of the *Encyclopaedia Britannica*, he waxed futuristic on Dancer's technique: 'Microscopic copies of dispatches and valuable papers and plans might be transmitted by post, and secrets might be placed in spaces not larger than a full stop or a small blot of ink.' While his latter reverie was to remain confined to the pages of spy novels, the former was genuinely prophetic: Brewster took examples of Dancer's work on his Continental tour in 1857 where they were seen by French photographer Prudent Dagron, who in 1870 used the method to relay messages by carrier pigeon between besieged Paris and Tours. In Italy, Brewster showed Dancer's slides to the Pope, as well as to the nobility of Florence. As quoted by Michael Hallet in the journal *History of Photography* (1986), Brewster reported:

The interest excited by these photographs was so great that I showed them to the distinguished jeweller Signor Fortunato Castellani, and suggested to him the idea of constructing broaches containing precious stones, so that the photographs may be placed within them, and magnified by one of the precious stones, or by colourless topaz or quartz formed into lenses.

Brewster's more academic interest in microphotography needs to be set within a broader context, one that takes shape by acknowledging that the principal scopic cultures of the age, or models for diverse ways of seeing, existed in states of conflict.

The scale of demand for microphotographs was the direct product of a rage for microscopy among the leisured classes. Increased availability of cheap microscopes, the mushrooming both in the capital and the provinces of scientific clubs and societies, the proliferation of scientific journals and magazines, as well as the new emphasis on leisure – sanctioned at the highest level with the restriction of the work day by the Ten Hours Act of 1847 – combined to elevate natural history to the science à la mode. Instrument makers up and down the country catered to an insatiable appetite for viewing such things as hogs' bristles, the antennae of gnats, the sting of a bee, spiders' mandibles, fleas' legs, the scales of eels, snowflakes and, strangely, the excrement of scallops. Dancer traded on the back of this craze for natural history. In his 1873 trade catalogue, a list of his microphotographs directly followed a long list of microscopic slides bearing a range of natural objects divided under the heads, 'Anatomical preparations', 'Urinary deposits', 'Insects mounted entire', 'Parts of insects', 'Vegetable preparations', 'Polarising objects', 'Sections of bones, teeth, minerals, fossils, shells, &c'. The implication was that the passage of the gaze from the productions of nature to the artefacts of light and collodion was a smooth one. In that the photograph had strong ties to a broader scopic culture that was saturated with various optical toys, illusions, magic lanterns, funny mirrors, dioramas, stereoscopes and kaleidoscopes, such a continuity of visual convention seems unlikely. The histology slide and the microphotograph brought different demands on the eye and had their roots in different specular practices. Natural history and histology concerned themselves with precision, calibration, veracity, with the domain of the real, while microphotography was intimately bound to the world of illusion, to multiplicity and reproduction, in other words, the domain of the displaced image. It was only by means of an ingenious, though probably unintentional, con, that microphotographs bridged the distance between the two scopic cultures of the day. To begin with, then, what kind of vision could the microscope lay claim to? And what characterized the scopic regime of natural history?

Before high-resolution achromatic microscopes became a common or garden household accessory, the investigation of nature had been the privilege of the scientific élite. But by the 1850s anyone with a microscope (and that, at least according to the rules of fashion, was everyone who was anyone) could replicate the findings of science, one of the principal pay-offs being the moral edification that attended a first-hand appreciation of the Argument from Design. For the genuinely curious, getting to know God the watchmaker was a lesser incentive than the possibility that through their microscopic activities they might extend the bounds of knowledge. Dozens of treatises on the microscope and its use appeared in the 1850s and 1860s, some catering for medical students, some for amateurs and yet others attempting to span the entire spectrum of microscopic utility, including the uses of microscopy in industry. Most paid homage to the advent of the achromatic microscope as the means by which the instrument was transformed from a toy into a research tool, so promoting microscopic research from an amusement to a science. Most made mention of reasonable price tags and reputable stockists of this most desirable piece of technology. And all spoke of the weird, intricate and wonderful world that sprang into existence because of it.

Tributes to the microscope flowed thick and fast. Edwin Lankester, editor of the *Quarterly Journal of Microscopy*, wrote in his *Half-Hours with the Microscope* (1859), 'what eyes would be to the man who is born blind, the Microscope is to the man who has eyes'. In *The Microscopist* (1851), Joseph Wythes MD, who recommended a mahogany-boxed £10 Dancer microscope, relied on a more traditional, gendered language of scientific prowess when he praised the microscope's ability to 'penetrate the arcana of nature'. The penetrative powers of the microscope as prosthetic eye were understandably revelled in, the implication being that instead of just seeing, the microscopist was enabled to *see into* and to *see through*. Vision that was thus enhanced, superhuman, was, imaginatively speaking, the nineteenth-century equivalent of X-ray specs, and microscopic literature was filled with talk of hidden worlds and unfathomed depths. In his *Evenings at the Microscope* (1859), the naturalist Philip Henry Gosse described in detail the way in which the microscope reveals an ordinary human hair to be an object of fascinating depth, comprised of successive layers of organic matter. A first examination yields a series of transverse lines apparently on the hair's surface; turn the screw and the lines grow dim particularly on the central part of the hair's cylinder; turn the screw again and another series of transverse lines becomes visible. Gosse explained: 'In fact, our eye has travelled, in this process, from the nearer surface of the hair, right through its transparent substance, to the farther surface.' William Carpenter, Professor of Physiology at the Royal Institution and of Forensic Medicine at University College, London, promoted the microscope in *The Microscope and its Revelations* (1856) as the best way to 'exercise the *observing* powers of a child' because it leads them to see 'how much there is *beneath the surface* even of what they seem to know best'. What they will find is 'inexhaustible life where all seems lifeless, ceaseless activity where all seems motionless, perpetual change where all seems inert'. Carpenter effectively implied that unaided vision was somehow defective, concealing a deeper truth about the world that only the microscope could reveal.

Carpenter, a prolific writer of scientific textbooks, was engaged in championing the microscope in a number of related debates. First there was the problem of rendering it a transparent mediator of nature. As far as access to nature was concerned, field naturalists had an obvious advantage over laboratory-based microscopists in dealing with nature *in situ*. Moreover, as a recent article by Graeme Gooday points out, they argued that the wrenching of nature into the laboratory or Victorian parlour – in essence a domestication of nature – involved the crime of distortion (*British Journal for the History of Science*, 1991). When Carpenter referred to microscopic study as 'healthful recreation', he was contributing to an ongoing debate about the relative merits of indoor versus outdoor microscopy, with the aim of proving the microscope to be as true to nature's multifarious productions indoors as out. By the end of the 1850s even the most outspoken critics of laboratory microscopy had been won over: Gosse, so effusive in his praise of the magnified hair filament in 1859, had in 1851, in *A Naturalist's Sojourn in Jamaica*, damned indoor microscopy as 'far too much a science of dead things; a necrology'.

Rescuing indoor microscopy from its tawdry reputation as a stultifying and unnatural closet activity brought Carpenter face to face with the related problem of making the field of nature unproblematically coextensive with the microscope's field of vision. That way laboratory or parlour microscopists could not be seen to be limited in their purview. The issue was as much a matter of semantics as anything else. Carpenter lamented the condition of the youth growing up in Britain's 'great towns' and 'vast Metropolis', 'whose range of vision is limited on every side by bricks and mortar'. For them the microscope represented 'almost the only means accessible under such circumstances' of tasting something 'of *life* or *reality*'. One senses that for Carpenter, who throughout the 1850s was developing the idea of mental physiology as a point of convergence between the sciences of mind and the natural sciences, the field of vision was as much a state of mind as a physiological circumstance, which was why extending it had such advantages as an educational aid, particularly for the poorer classes. It taught them how to think, and also how to think better of themselves. Carpenter reached giddy heights of utopianism with his conviction that microscopic study raised the 'labouring population' from 'the grovelling sensuality in which it too frequently loses itself'. Invoking a deliberate slippage between the natural and social orders, he wrote, 'for we cannot long scrutinize the "world of small" ... without having the conviction forced upon us, that size is but relative, and that mass has nothing to do with real grandeur.'

The perception of the microscope as a democratizing instrument that empowered the labouring masses at the same time as educating them had widespread appeal to a liberal bourgeoisie who unhesitatingly identified the universalization of their values with progress. In his 1859 treatise on microscopy, Gosse claimed that the student who shall have verified for himself the observations here detailed, will be no longer a tyro in microscopic science, and will be well prepared to extend his independent researches, without any other limit than that which the finite, though vast, sphere of study itself presents to him.

A proper training of the observing faculties would lead to an emancipation of spirit, just as enlargement would lead to enlightenment. An article on 'The Microscope in Education' in an 1868 issue of the *Student and Intellectual Observer* was even more direct on this count, appearing as it did just a year after the Reform Act enfranchized some one million artisans and members of the urban lower middle classes. The article aligned itself with those 'with an eye to political interests, and the bearings of democratic change' who 'consider that national safety depends on national education'. And the programme of national education it recommended was one that gave 'preference to plans which bring truths home to the eye, and with this view it is impossible not to regard the microscope as one of the foremost instruments for the communication of knowledge'. The microscope could be regarded as a socially responsible tool only because it was seen to be unfailing in its ability to deliver truth; as such, it would steer the lower classes on a proper course of rational self-improvement, not distract them with chimeras.

In the new positivist cosmos where truth was identified with facts, with discrete units of unambiguous knowledge, microscopic information (like all raw scientific data) required careful handling. It needed to be marketed as evidence that spoke for itself, that is to say, as evidence which short-circuited the troublesome tangle of interpretation. The enterprise was of the order of a public-relations exercise, involving airbrushing into invisibility the very processes that manufacture meaning. The greatest test of the microscope's ability to deliver truth lay in calibrating what was viewed under the microscope relative to what could be seen with the naked eye. The formerly invisible world of the minuscule had to be measured in order to be believed. Seeing alone could not command credibility. Such an epistemological shift effectively marked the disappearance of the scientist: belief, which expresses a relation between the observer and whatever is being observed, was replaced by credibility, a property residing entirely within the object.

For the benefit of amateurs, a preoccupation with calibration was translated into counting. Gosse, for example, in his ode to the human hair, provided the following information: that a hair of 1/10th of a line thickness contains roughly 250 fibrils in its diameter and 50,000 in its entire calibre. He also reported that 24,000 convex lenses were contained in the two eyes of large species of dragonfly. As if to militate against his readers understanding numbers as mere titillation, he went on to point out the commercial uses of counting, by explaining that the excellence of wool related directly to the closeness of the imbrications of its fibres. Merino wool has 2,400 serratures per inch, Saxon wool has 2,720 while Leicester wool has only 1,850 – the greater the number of serratures, the better the wool's felting quality. Microscopists addressing their peers took measurement far more seriously. The journals of the Queckett Microscopical Club and the Microscopical Society were filled with notices of new micrometers, and articles debating whether more accurate measurements might be made with platinum-wire micrometers, mechanical micrometers or glass-engraved micrometers, with micrometers mounted on the microscope's stage, or cycpiccc, or both. The more precise the measurement, the more credible the object.

The making of ever more accurate micrometers was a twin preoccupation. In his article on the micrometer for the *Encyclopaedia Britannica*, Brewster congratulated Sir John Barton for succeeding in using a diamond point to engrave divisions of 1/500th to 1/10,000th of an inch on glass. By taking impressions of this on transparent films of gelatin, a very good micrometer could be obtained. Dancer too would no doubt have taken his hat off to Barton. He had long been engaged in the precision calibration of scientific instruments. Indeed, in 1843 he had constructed the thermometer that James Joule used in his famous experiment determining the mechanical equivalent of heat, announced in the *Philosophical Magazine* in 1845. Brewster went on in his article to pay tribute to Dancer, who, in his opinion, had come up with the means of making even more accurate gratings. He enthused that in Dancer's micro-photographs the film of collodion 'is so thin and transparent that it is invisible, and allows objects to be seen through it as distinctly as if it were the thinnest glass'. The appeal of this was irresistible: 'If a system of opaque or transparent lines therefore is impressed upon it photographically, when reduced to the minutest size, for a system of large and sharply defined lines, we shall have the most perfect micrometrical scale that can be conceived.' Brewster had written to Dancer requesting him to make such a micrometer in July 1857 and a letter dating from September that year thanks Dancer for complying, adding, 'I cannot doubt that they will be superior to other gratings.' Dancer makes brief mention of the fact that he pro-duced photomicrometers and diffraction gratings commercially, pre-sumably from this time onward.

Calibration, however, was not in itself sufficient to overcome all the obstacles in the way of validating the 'world of small'. Corrobora-tion was required. Only then could the microscopist's subjectivity be discounted. The problem was succinctly put by Carpenter when he expressed concern that the observer at a microscope was liable to record 'not what he sees in it, but what he fancies he can see'. Although a rigorous training in the techniques of scientific observa-tion, a learning of guidelines as it were, went a long way towards keeping the imagination in check, vision still needed to be standard-ized. The French instrument maker Nachet hit on the ingenious idea of developing a microscope at which two, three or four people could observe simultaneously. His instrument, however, was more significant in encapsulating the problem than solving it – scientific communities did not often convene in a single city, let alone around a microscope tube. Corroboration had to function effectively across continents and down the years if microscopic facts were to have a universal purchase on truth. Much of the attention to the issue of standardization in the various microscope handbooks therefore con-centrated on the production of drawings. These, Gosse insisted, must be 'accurate delineations of the objects represented'. The use of micrometers was deemed essential, ensuring that drawings made by independent observers could be calibrated in terms of scale, but

what was open to discussion was whether drawings should be out-
lines or details, whether they should be produced with artistic freedom or with the aid of a camera obscura, and whether they were
best reproduced by lithograph, woodcut, copper plate or engravings
on stone.

Lionel Beale, Professor of Physiology at King's College, surveyed
the whole question of sketching microscopic observations in *How to
Work with the Microscope* (1857), finishing up by assessing the value
of photography. He began by stating that the merits of photography
in reproducing microscopic observations 'are so obvious that I need
not occupy your time in recounting them'. He was of course referring to the photograph's fidelity to the real. He then described how
photomicrographs could be obtained by simply removing the eye-
piece of the microscope and attaching a camera in its place. The
mechanics might have been straightforward, but decent results
proved extremely hard to come by. The main problem was that 'the
foci of the chemical and visual rays are not coincident in the ordinary object-glasses, consequently several experiments have to be
made in order to find the exact focus of the chemical rays, and we
must take different pictures until one is found to be in focus'. Haphazard and fiddly, the process was further encumbered by photography's inability to negotiate depth; if a tissue specimen was too thick
or opaque, photomicrometry could not penetrate it, and because
the taking of a photograph demanded a decent quantity of light,
such depth of focus as existed was undermined. Moreover, circulat-
ing illustrated papers was hampered by the fact that reproducing
such images as albumen prints was prohibitively expensive. Photo-
micrographers were admired for their dexterity and dedication but
in the 1850s and 1860s they remained a small enclave of enthusi-
asts within the microscopical community. Dancer, ever eager to
turn his hand to practical problems, contributed to the genre by
devising a diaphragm for the stage aperture that when contracted
would increase the depth of focus of the microscope without dimin-
ishing illumination. In his own inimitable style he failed to patent
this device, which was thereafter known as the Davis shutter.

This then was the scopic culture of microscopy: preoccupied with seeing into and through things, with defining the field of vision, with depth, accuracy, measurement and corroboration. How different it seemed from the free-wheeling sensorium that was the world of optic spectacle. So alluring was the culture of illusion that even the natural history contingent could not resist an excursion into it. In her *Sketches with a Microscope* (1857), the popular science writer Mary Ward talked of the 'microscope's magic tube', while Gosse employed an even more outlandish allusion to the supernatural. In *Evenings at the Microscope* he wrote: 'like the work of some mighty genie of Oriental Fable, the brazen tube is the key that unlocks a world of wonder and beauty before invisible, which one who has once gazed upon it can never forget'. This was the language of the laboratory leaning on the language of the fairground. Joseph Wythes, another proselytizer of serious-minded microscopy, was similarly drawn to the glitzy, tricksy culture of scopic spectacle. He took the greatest pleasure, for example, in describing an experiment with a fly's eye containing in excess of four thousand lenses; in an 'observer who looks through it at a distant candle &c, the interference of the light in the minute lenses will cause a number of images to be perceived, tinged with beautiful colours'. Elsewhere he claimed that 'the splendid colours and systems of coloured images produced by transmitting light through transparent bodies that possess double refraction are the most brilliant phenomena that can be exhibited'. Clearly, the new apologists for the microscope had trouble leaving the kaleidoscope in the nursery.

J. B. Dancer was himself steeped in the culture of spectacle. As a boy he had helped his father stage outdoor public exhibitions with the solar microscope, a sort of lantern-projection microscope that enabled images of small natural objects to be directly screened on to a wall. Josiah Dancer had constructed a particularly large version of this instrument that used a twelve-inch condensing lens. Its effect was magnetic, drawing crowds to gasp and gawk at the images of human hair magnified to six feet in diameter and cheese mites grown to four or five feet in length. It has already been noted that Dancer put his own showmanship into practice in 1840 when he wowed a Liverpool audience by producing a Daguerreotype of a

flea. Spectacle and illusion marched hand in hand through these
decades – the magic lantern and the stereoscope being perhaps the
most popular devices that welded the act of vision to an ethos of
deception. Used initially to thrill children by conjuring up shadows,
ghouls and phantoms in their living rooms, the magic lantern
became a tool of public education. In 1837 John Smith, proprietor
of the *Liverpool Mercury*, asked Dancer to improve the mode of illumination in the lantern so that he might illustrate his geography lectures in a large hall. The old argand oil-lamp could only produce an
illuminated disc of seven feet in diameter, not bright enough to be
seen distinctly from any distance. Dancer sucessfully adapted the
Drummond light (better known as limelight), obtaining a radiant
fifteen-foot disc. Not one to let a technological possibility slip, he
capitalized on the gas operation of limelight and came up with the
Dissolving View Lantern which displayed one scene seductively melting into another and which soon outstripped the popularity of its
predecessor. Ever more sophisticated versions of this hardware left
Dancer's workshop, his pride and joy being the wonderfully named
Diagonal Bi-unial Lantern which *Chadwick's Lantern Manual* described as 'one of the most perfect of its day'. Lantern exhibitions
held regularly at the Manchester Mechanics Institute and elsewhere
became a lucrative sideline, though Dancer probably took greater
relish in the opportunity such events offered for indulging what can
only be described as his inventive compulsion than in lining his
pockets. When the Manchester Institute extended its entertainment
programme to include annual winter exhibitions, his Fairy Fountain
became one of the chief attractions. This optical extravaganza,
which consisted of a huge multi-jet fountain, illuminated from below
by coloured lights activated by an electric keyboard, sent a reviewer
from the *Manchester Guardian* into raptures: 'the wonderful, glitter-
ing sheen of that magic and myriad tinted fountain, scattering its
thousand liquid rubies and emeralds, its topazes and amethysts, its
iris glories and gold and silver jets, dazzles and delights all eyes with
a new sense of the beauties of colour and motion' (quoted by L. L.
Ardern in *John Benjamin Dancer*, 1960).

The reviewer's linking of multiple visual sensations to pleasure is telling: delighting, dazzling and pleasuring the eyes was the simple, though gargantuan, aim of all this spectacle. Sensual gratification as a specular goal may be read as a mode of Romantic self-absorption, transporting the observer from external reality into the realm of the senses, so that he might acquaint himself with and marvel at the body's mechanisms of perception. This kind of subjective vision turned on the productivity of the viewer. Like Goethe and Schopenhauer who were so taken with the phenomenon of retinal afterimages, or Brewster, who was fascinated with how the 'brain unites' two different pictures seen through a stereoscope 'into a single apparently solid reconstruction of the original', Victorian consumers of optical toys and gadgets were embarking on an inward journey, exploring the creative self and the nature of perception. The narcissism inherent in the enterprise is neatly symbolized by the phenakistiscope (literally 'deceptive view') invented by Belgian scientist Joseph Plateau in the 1830s. This amusement, predicated on the afterimage, demanded that viewers position themselves before a *mirror* and peer through slits in a spinning cardboard disc in order to witness images of a girl skipping or a horse trotting painted on the reverse of the disc come to life in illusory motion. There is an undeniable peephole mentality at play in the most innocent specular activities, a private, perhaps even illicit, pleasure in experiencing the heady whirls of kaleidoscopic permutations, the teasing visual sequences of the phenakistiscope, or the stereoscope's power to make flat images loom out in fleshy fullness. It was this almost orgasmic excess that prompted Baudelaire to write, in a famous passage from 'The Modern Public and Photography' which describes the salons of 1859, of 'greedy eyes . . . glued to the peephole of the stereoscope, as though they were the skylights of the infinite'. A sense of disorientation, such as was also provided by halls of mirrors and dioramas, further compounded the way in which observers were thrown back on themselves in order to find some point of anchorage in the real. This private knowing was essentially symbolic or notional, having more to do with the pursuit of the viewer's stable identity than with the fixity of the object in view. As such it was a world apart from microscopical visual lore. There is some irony in the culture of public spectacle being productive of private knowledge, while the solitary activity of the parlour microscopist revolved around public knowledge and consensus.

The advent of photography was a gift to the culture of spectacle. For here was a means of producing images whose existence was totally independent of their original referent, and of producing them again and again and again. A dizzying profusion of displaced images thus entered the world of the Victorians, probably engendering, as does today's virtual-reality industry, a terror of information overload and the multiplication of endless possibilities. Photography, moreover, had no primitive stage; it sprang on to the scene as a full-blown science and its early development was a matter of extending not its technology but its scope. Almost as soon as it appeared, photographic glass-plate images were made into lantern slides; there was a proliferation of photographic panoramas and dioramas; instrument makers' catalogues carried long lists of photographic stereoscopic views. And there was an exponential growth in the number of its practitioners; the 1851 census for Great Britain identified 51 persons as photographers by occupation; in 1861 there were 2,879. Most of these would have been studio photographers thriving on the fashion for portraits and *cartes-de-visite*.

Alongside the growth of the new profession, there existed a race for 'firsts': in 1852 Warren de la Rue in London became the first person to obtain photographs of the moon, Porro and Quinet in Paris recorded the solar eclipse of 1858, while London-based Antoine Claudet used his photographometer in attempting to measure the varying intensities of light rays. In 1858 also Nadar produced the first photograph to be taken from a tethered balloon, and in 1861 Colonel Aimé Laussedat manufactured the first map based on aerial photography. A sense of photography's unlimited scope additionally caught the imagination of Lewis Carroll, who became one of the century's most distinguished photographers. In the same year that Carroll began his love affair with photography, 1855, he wrote the short story 'Photography Extraordinary' for the *Comic Times* in which he fantasized that the camera could penetrate a sitter's mind. He mused, 'The ideas of the feeblest intellect, when once received on properly prepared paper, could be "developed" up to any required degree of intensity'. The process, he explained, depended on 'a mesmeric rapport between the mind of the patient and the object-glass'. An unsuspecting guinea pig, who claimed to be thinking of nothing, yielded a photograph covered with writing of a miserable quality, belonging to the 'milk-and-water School of Novels'. The next stage of development added colour and atmosphere, moving it to the 'strong-minded or Matter-of-Fact School', while, with the final degree of development – a garnishing of storm and stress – it entered the ranks of the 'Spasmodic or German School'. Various experiments followed, like 'working up a passage of Wordsworth into strong, sterling poetry'. In wry spirits, Carroll relates that 'the same experiment was tried on a passage of Byron ... but the paper came out scorched and blistered all over by the fiery epithets thus produced'. The story ends with Carroll archly hoping against hope that this evolutionary art might successfully be applied to parliamentary speeches. Photography made deep and lasting impressions on Carroll's own psyche, and it is tempting to see photography as providing the leading metaphor for the parallel worlds of the Alice books, given that it reverses images like a mirror via a positive–negative correspondence. But it is even more tempting to see in microphotography proof of a real looking-glass world, where landscapes, buildings, animals and people expand and shrink on whim.

Microphotographs were tailor-made for the 'greedy eyes' that feasted on the culture of spectacle. Like the consumer of stereoscope or lantern slides, an observer could, within a matter of minutes, gallop through a succession of images, from a photograph of a Landseer, the moon, the Great Pyramid at Giza, the frontispiece of the *Illustrated London News*, to a portrait of Garibaldi, obtaining a flow of visual stimulation more or less on tap. But, unlike these consumers, the microphotographic observer could summon whole worlds into being from nothing – if not an act of creation, than at least a process of discovery. The sense of discovery invoked was, moreover, a scientific one, in that the necessity of using a microscope entailed a borrowing of the scopic régimes of natural history. It was this collision of specular cultures in microphotographic recreation, a mixture of rational

and irrational viewing paradigms, that set microphotographs apart
from other optical toys. Like natural historians, microphotographic
observers could broaden their horizons by staring down a brass
tube, extending their field of vision so that it encompassed all there
was to see. In that there existed a correspondence between visual
fields and states of mind, there are overtones of imperialism here,
for, like the Great Exhibition of 1851, which garnered the whole
world under one roof, microphotographs delivered the world on a
pinhead. Perhaps, too, in this stage-managed 'miracle', there is some
secular riposte to the old and trivial scholastic debate as to how
many angels were able to dance on the head of a pin. At root, what
microphotographs were pandering to was a desire for empowerment
and containment, a totalizing vision that exceeded anything offered
by panoramas and dioramas, that was, in effect, encyclopaedic.

Many of Dancer's microphotographs manifest the obsessions of
the microscopist figuratively. Brewster had written in the *Encyclopaedia Britannica*, 'It has been a long trial of skill to include the Lord's
Prayer in the smallest circle by the unaided hand of the writer', as a
prelude to discussing the efforts of Sir John Barton and of M. For-
ment in Paris who managed to compress a piece of writing of 3
2/10 inches diameter into the space of 1/30th of an inch. Dancer's
technique accomplished the task effortlessly. In fact, tiny, handwrit-
ten copies of the Lord's Prayer were popular pieces of *virtu* in the
Renaissance, a sort of 'European netsuke', a luxury whose manufacture Dancer mechanically transformed into routine. Echoing the
natural historian's preoccupation with counting, while also betraying a faint air of smugness, his various slides of the Lord's Prayer
bear the labels 'Contains 280 letters', while the label affixed to Lord
Raglan's tablet boasts 'Contains 1,687 letters'. Dancer also produced
numerous group portraits in which a sea of faces swarm within an
oval or circular frame. With titles like 'Ninety-seven Dramatic Portraits', '112 Portraits of Eminent Men' and '155 Portraits of Eminent
Persons', these were obviously designed not merely to delight the
eyes, but, to use Carpenter's term, to exercise them. The active
viewer was being encouraged to identify the faces and to count. But
even more revealing are a number of slides of an entirely self-referen-

'112 Portraits of Eminent Men'. Slide no. 149 from Dancer's 1873 catalogue:
Shakespeare, Peel, Napoleon, Luther, Newton, Byron, Wren, Dickens,
Handel and Titian cheek by jowl with dozens of other glitterati.

tial nature that address microscopic vision – 'The Lord's Prayer in a Pinhole', 'The Eleventh Commandment in the eye of a needle', 'The National Anthem, contained in the eye of a needle' and so on. Although such images represent the microscopist's concern with scale, they do so in the abstract, gesturally. There is no actual need for measurement, since the minuscule image is in every way identical to its real-life original. By virtue of their verisimilitude, micro-photographs assimilated the process of calibration to the act of recognition. James Nicholls claimed in his book on the mysterious art of microphotography that it was possible to produce 'a photo-graph of a giant cathedral wherein you may count every stone, and tell the hour by its great clock'. His brag was about how accurate microphotography was in the delineation of detail: you can count, but the point is, you don't have to. Because of photography's much-vaunted fidelity to the real, you can take it on trust.

Herein lies the false promise of microphotography. It offered con-sumers the illusion of being a scientist (the cultural heroes of the hour). They were compelled to follow the procedures of natural history, peering down a microscope tube, fiddling with the focus and bringing an object into perfect view. They were invited to count, to appreciate the importance of measurement and obtain some notion of scale. Not only were consumers of microphotographs able to believe that, like scientists, they could make valid judgements about the world, but they could do so without the rigours of calibra-tion. This is why there are no pictures of fleas' legs, hogs' bristles, or spiders' mandibles. The inclusion of natural-history slides would have given the game away, exposed the deception.

That Victorian adults should take such childish delight in using photographic miniatures to play at microscopy sits oddly with the dominant image of bridge builders and explorers. Yet there was nothing regressive about their attempted return to the nursery. Microphotographic recreation seems rather to have been part of a collective idealization of childhood, which included the idea that in order to move forward it was first necessary to move back. Words-worth's notion that 'The child is father of the man' aptly describes the reversal of the learning curve that characterized the quest for empirical knowledge. The child's mind, refreshingly free of theories, prejudices and learnt conventions, was the best model of the neutral observer. Ruskin, for example, in 1857, wrote in *The Elements of Drawing* that

the whole technical power of painting depends on our recovery of what may be called the innocence of the eye, that is to say, of a sort of childish perception of these flat stains of colour, merely as such, without conscious-ness of what they signify, – as a blind man would see them if suddenly gifted with sight.

Edwin Lankester had also invoked the blind man to drive home the point about scientists viewing the world *de novo*, dew-fresh, as if for the first time. It was the innocent eye that was best qualified to broach the new and the unknown with a view to obtaining truth.

Innocence, however, can only thrive in a safe world. Destabilize the world and the innocent flounders, becoming a victim. Micro-photography belonged to the world before Pasteur, before the very small threatened danger, disease and death, and the innocence it perpetuated was that the minute presented only differences in scale not of kind. The microphotographic universe, though elastic, was fundamentally stable; within it things could expand or shrink, but they preserved their identity. Its effect on the observer was therefore reassuring. This is what Carroll's Alice learns from her adventures – however small or big she is, she is still Alice; even when she is growing in the house she retains her composure. Not even the pur-suit of divisibility seemed to threaten established ideas about integ-rity; the cell was regarded as a self-regulating unit, while the atom was understood as the smallest particle to retain the chemical proper-ties of a substance. For all the sensory excitement it elicited, mechani-cal minuteness, the product of either miniaturization or division, kept the world the same. Biological minuteness proved to be of a different order. Initially, minuscule living organisms were assimilated into the world of design. In 1859 Gosse delighted in the elegance, delicacy and transparency of the 'invisible animalcules' collectively known as *Infusoria*, which could be procured by simply steeping vegetable matter in water for several days. Even though Theodor

Schwann had linked germ theory to disease in the 1830s, most scientists persisted in thinking it absurd to suppose that tiny, living agents could invade and kill vastly larger organisms. It took Pasteur's work on fermentation to open the door to the exteriority of disease. By the 1870s, when the popularity of microphotography was on the wane, the world of small had become a hostile place. Its fate as adversary was sealed in 1877 by Pasteur's paper on the bacterial causes of anthrax. Thereafter, the biologically minute became the invisible enemy, breeding paranoia, fear of contamination and the desire for immunization. If the microphotograph encapsulated our relationship to the world of small before the acceptance of germ theory, its successor was the vaccine.

Transparency: Towards a Poetics of Glass in the Nineteenth Century

ISOBEL ARMSTRONG

I

A superb looking-glass curtain, which drew up and let down in the sight of the audience, and reflected every Form and Face in the gorgeous house.
James Fitzball, *Thirty Five Years of a Dramatic Author's Life*, 1859

After an overture, to which no attention of course was paid by the excited and impatient audience, the promised novelty was duly displayed; not one entire plate of glass – that could not have been expected – but composed of a considerable number of moderately sized plates – I have seen larger in some shop windows – within an elaborately gilt frame. The effect was anything but agreeable. The glass was all over finger or other marks, and dimly reflected the two tiers of boxes and their occupants. It was no imposition, however, it was a large mass of plate-glass, and in those days must have cost a great deal of money. There was consequently considerable applause at its appearance. The moment it ceased, someone in the gallery, possessing a stentorian voice, called out, 'That's all werry well! Now show us summat else!' What more cutting comment could the keenest wit have made upon this costly folly? Did the manager who was guilty of it deserve to succeed?
Lee J. R. Planché, *Recollections and Reflections*, 1872

So the 'looking-glass curtain' at the Coberg Theatre (now the Old Vic), a sensation of the 1830s which weighed five tons, was not made of mirror glass. This is a moment when two readings of glass converge and conflict. One relies on an ancient way of thinking of the body reflected from an external surface, the mirror: the other responds to the new production of mass-produced transparency, in which one's body can be, glancingly, inadvertently, reflected back from the environment, belonging to the urban phantasmagoria outside one's control. For the first time in our culture, perhaps, the self can be a mirage returned from the surfaces of the city landscape.

Once the idea of the mirror is relinquished, glass is confounding. For the first writer the mirror includes everyone in the theatre, 'every Form and Face', from top to bottom, unproblematically reflecting social unity in the 'gorgeous house' at a time when British society was actually highly stratified – and marked by social division. For Planché, on the other hand, the genre of this finger-smeared plate glass is uncertain. With its elaborate gilt frame the analogy is the aesthetic one of the portrait, the individual subject gazing from a wall with which the spectator communes. But it is also that of the commercial plate-glass shop window – 'I have seen larger in some shop windows' – then expensively produced but increasingly common. The spectator is a consumer gazing at commodities which awaken desire more effectively than the 'dimly reflected' tiers of boxes and their occupants. Do you, then, confronted with these strangely organized squares of 'moderately sized plates' of glass, look on it, through it, at it? And what of the reflected audience? For those who are *in* the boxes will see themselves, however dimly, but other people will see different groups of people from different angles in the theatre, depending on where they are. The illusion of collective seeing enabled by the proscenium arch is fractured, splintered into individual acts of seeing: there is a problem in locating the self in relation to others, others in relation to self, in or on this shadowy screen; who sees what? who sees you? at what angle?

Two understandings of spectacle intersect – the proscenium arch and the screen, the simulacra of the screen superimposed on an older interactive genre of gesture and declamation. The 'stentorian voice' calling for something else is not calling for the play to begin but following through the logic of static images on a screen, that they should be succeeded by other images, that they should begin to move. The 'looking-glass curtain' with its ambiguities interposes itself when mass-produced glass, *the* innovative material of the nine-

teenth century, was beginning to make possible new cultural meanings and opening up cognitive space, transforming consciousness.

The beginnings of an avidly scopic culture – a culture of *looking* –
are marked in the Coberg curtain. Glass, in the endless multiplicity
of forms capable of being generated from just four elements, prefabricated panels, the lens, windows, containers, was always in between.
Its fine gradations of invisibility provoked, if not a crisis of mediation,
acute awareness of an intensely mediated culture, whether you
looked at the uncertain meaning of the heavy squares of Coberg
plate glass, down a microscope, or at the optical illusions created
through glass in a London pleasure garden. An earlier culture's
experiences can become metaphors for subsequent knowledge.
Both Freud and Lacan *read* glass, reusing nineteeth-century dis-
courses, understanding the immanent meaning of transparency. In
1900, in *The Interpretation of Dreams*, Freud wrote:

We should picture the instrument which carries out our mental functions as
resembling a compound microscope or a photographic apparatus or something of the kind. On that basis psychical locality will correspond to a point
inside the apparatus at which one of the preliminary stages of an image
comes into being. In the microscope and telescope as we know, these occur
in part at ideal points, regions in which no tangible component of the
apparatus is situated.

In the first seminar in *Écrits* (1975), Lacan says:

I cannot urge you too strongly to meditate on the science of optics ...
peculiar in that it attempts by means of instruments to produce that strange
phenomenon known as *images*, unlike the other sciences which carry out on
nature a division, a dissection, an anatomical breakdown.

Freud saw that glass is a philosophical material, not only because
it formed the substance of what were called in the nineteenth century 'philosophical instruments', the microscope and telescope, the
apparatus constructed for the pursuit of 'natural philosophy', or the
sciences, but because through its mediation ideal images were
formed, created by means of matter but not coincident with it. Lacan
follows this through, contrasting the sciences which work with a

kind of brute literalness on matter, or nature, itself, with optics, whose object is to use matter to create immaterial phenomena, images – we can photograph the illusory rainbow. The sheer abstractness of glass becomes a way of representing the displacement of consciousness, or rather, of consciousness as the instrument for creating specularity and the simultaneous displacement of the 'ideal' specular image.

In some preliminary observations by Sir John Herschel on the telescope in 1861, everything noticed by Freud and Lacan, along with most of the components of the culture of looking made possible by glass, are present. If glass is placed before a lighted object, either as a refracting or reflecting surface, a picture of the object is formed at a certain distance behind or before the glass surface. This picture is distinct only at one particular distance from the glass surface, the focal distance. The picture can be viewed on the same side as it is formed, thrown on to a panel, magnified, and inspected by numbers of people: or it can be thrown on to a screen of roughened glass and viewed from the hinder side. But in these cases it is not really a telescope but a camera obscura, sharing with the image 'fixed photographically, and thus, becoming a real object [that] may be preserved,' the characteristic of revealing minutiae not evident to the naked eye in 'the real object'. The telescope, on the other hand, like the microscope, is an instrument for transferring the image direct to the retina, 'larger and clearer' than it would be without its mediation: it is an *individual, private* instrument. Herschel adds: 'An imaginary picture, or what in optics is called an *image*, is, or may be conceived to be, formed in the air, but it is not visible *as a thing* to an eye situated out of the direction of the rays which go to form it.'

The ideal space of the focal image, the 'real' object, the politics of private viewing, the reflecting and refracting powers of glass, all these, with the huge expansion of mass transparency, became cultural as well as scientific *experiences*, and thus new kinds of problem as they were transposed into new sites of production. Technologies originating at the Chance factory in Birmingham in the early 1830s, the discovery of fresh ways of constructing massive glass and iron structures, the abolition of the glass tax in 1845 (so that its price fell from 4s 6d per foot for the best glass at the start of the century to under 2½d) meant that mass production transposed élite forms and practices to street and urban spectacle. Mass production reduplicated the glass room or greenhouse in innumerable ways: it mimicked the instruments of high research in *trompe-l'oeil* gadgets and optical toys, using scientific principles with wanton prodigality in spectacle. The virtuosity, the *jouissance* of glass, culminated in the twenty-seven-foot-high glass fountain of the Great Exhibition of 1851, where the static transparency formed by technology and the moving transparency of water converged.

There is a perfervid intensity about Dickens's celebration of the making of plate glass in a 'hall of furnaces' under the new conditions of free trade, as described in an article he wrote with W. H. Wills in *Household Words*, 1851. The works he visited produced as much plate glass in a week after the removal of excise as all the manufactuaries in England before its removal. Now this 'diabolical cookery' could be conducted by 'knots of swarthy muscular men' in the unfettered freedom of 'unrestricted' trade. The 'salamander' in 'human form' could sit triumphant on his trolley, balancing a cauldron of red-hot glass by playing see-saw with it.

The dreadful pot is lifted by the crane. It is poised immediately over the table; a workman tilts it; and out pours a cataract of molten opal which spreads itself, deliberately, like infernal sweet-stuff, over the iron table; which is spilled and slopped about, in a crowd of men, and touches nobody. 'And has touched nobody since last year, when one poor fellow got the large shoes he wore, filled with white-hot glass.'

II

Transparency. Limpid, translucent, *invisible*. Only a gradation of opacity, a sheen of light, makes one aware of it. 'The limits of the diaphane ... adiaphane in': the limits of the thing you can see through. Joyce's Stephen Dedalus, in *Ulysses* (1922), realized the paradoxes of glass's nature when he asked not only where the transparent object could be said to begin or end but what limits there were to transparency, what residue of visibility in the medium itself. What was not visible was the labour that created it. 'Poor fellow.'

III

If glass was blown, and in mid-century blown sheet glass, extended to four-foot cylinders, cut and rolled, became the commonest form of manufacture, then the observer would look through the residues of somebody else's invisible breath. Look through in two senses: one would look past, and look by means of, the breath which had distended red-hot glass and which was now sealed within it. The plate glass Dickens saw was cast, rolled, annealed, steam-ground, by groups of 'swarthy' workmen, smoothed by women (whose ceaselessly stretching bodies he found erotically fascinating), and finally completed in a rouging process operated by young boys, in a room 'glowing with red', incarnadine. He is nonchalant about the horrors of the processes necessitated by white heat – 'one poor fellow got the large shoes he wore, filled with white-hot glass' – a mutilation occasioning a pun defamiliarizing glass slippers. The labours of these workmen, sipping their employer-supplied beer outside the factory, no longer the infernal creatures of a daemonic furnace, must have entered the optical unconscious (the phrase is Walter Benjamin's) of the perceiver. They must have entered the structure of artefacts and texts. But no one knows what the 'swarthy' workmen thought of their visitors.

Ruskin thought of the incessant motion of glass bead makers, cutting up rods of glass, as the very type of the division of labour. In the famous chapter of *Stones of Venice* (1853) on the nature of Gothic, he wrote of 'the men who chop up the rods sit at work all day, their hands vibrating with a perpetual and exquisitely timed palsy, and the beads dropping beneath their vibrations like hail'. He believed that aesthetic objects in particular must be marked by the division of labour, because a society's forms of play, to which the aesthetic is related, are structurally related to its form of work. But whereas 'human labour' is 'legibly expressed' for ever on other artefacts, glass, exasperatingly, erases this connection, its own invisibility making the conditions of its production invisible.

How does a poetics of transparency redeem, or recognize, this
invisibility? How does it recognize that glass meant different things
to different groups? If spectacles, as Regenier Gagnier has pointed
out, meant knowledge, intellectual empowerment and freedom to
the middle-class reader, they meant the threat of unemployment to
the working class or artisans if the master saw a man whose eyes
were failing.

A poetics of transparency will be involuntarily beautiful – too
beautiful, perhaps, in representing its object. It will see, as Christina
Rossetti saw in old glass, in *Time Flies* (1885), 'Such a contour, a
curve, an attitude if I may so call it . . . as a petrified blossom bell
might retain, or as flexibility itself or motion might show forth if
these could be embodied and arrested.' She added, 'Inert glass
moulded from within caught the semblance of such an alien grace.'
Alienated labour becomes 'alien grace', and the glass blower's exhalation shaping 'inert' matter from within, arguably a violence done
to artisan and material, becomes the unique fusion of soul and body,
breath and matter, a form made by a God who can penetrate to the
innermost spaces of the subject, creating from the inside out.
Agency, not powerlessness, belongs to the glass blower. But as if
recognizing that this beauty is won too easily, Christina Rossetti
contemplates the effects of oxydization, a filmy iridescence which is
the result of chemical reaction in the environment and the effects of
time, acting on the outside of glass, not from 'within', and offers a
materialist history. 'I one day rescued from an English roadside ditch
a broken bottle: and it too was oxydised! . . . it too displayed a
variety of iridescent tints, a sort of dull rainbow.' Common bottles
were made from moulds by that time, not shaped by a 'breath or a
blast from within'.

The moral point, that the 'many' can be 'thankful for dim rainbows', and its strikingly democratic political reading are less arresting, perhaps, than the form of her meditation. Brief, fragmentary,
discrete passages simply juxtapose the high aesthetic artefacts
of old cultures, of Rome and Venice, with the common ditch

bottle, making no attempt to force a false reconciliation through comparison, but by this very juxtaposition demonstrating the difficulty of the relationship. This tentative effecting of typology through withholding has something of the power of Walter Benjamin's 'dialectical image', blasted out of the continuum of history to produce a shocking and problematical resonance with the present – not surprising, as Susan Buck-Morss shows in her work on him, *The Dialectic of Seeing* (1991), since Benjamin was creating a kind of surrealist typology. But Rossetti stays with the difficulties, the particulars which put the dialectic under strain: 'For *in a minor key* [emphasis added] it too displayed a variety of iridescent tints.' The dispossessed are not here simply added in conscientiously to complete the social analysis: there is no attempt to suggest a unitary culture capable of transcending division. The same processes, different kinds of agency. This suggests that a poetics of transparency can only theorize the culture of glass and 'rescue', as Rossetti put it, the labour it conceals, by staying with just these difficulties, worrying with their particularities. And it may be that the most helpful way of doing this is by adopting Rossetti's own method of discrete, unselfsufficient fragments of lyrical thinking.

IV

Inert glass moulded from within caught the semblance of such an alien grace.

V

The discontinuities, then, the mode of lyrical thinking, draw attention to disconnection, to a difficult dialectic.

But why mediation?

To see through somebody else's breath. 'Thought through my eyes . . . the limits of the diaphane': the diaphanous or transparent object. Stephen Dedalus speculates that thought occurs through the eye's mediation: thought occurs by means of and because it passes through the transparent lens of the interpreting eye as data is collected by the eye: what comes in as perception flows out as thought. The mediation of glass, an external or prosthetic eye (window comes from Old Norse 'wind's eye'), troubling because, as with the eye, you cannot see yourself seeing, is also troubling because you *can* see *it*, the fragile interposing sheet of almost nothingness with a sheen to it. As transparency mediates the gaze, its lucidity is itself the result of a highly mediated process. Glass is a *transitive* material, and in the nineteenth century it became insistently so, a material which made one intensely aware that, in the movement between one state and another, there is a third term. The world is triangulated. From seeing to not seeing in the dark – the telescope can show the dial hands of a clock miles away in the dusk, Herschel said: from looking inside to looking outside – the cumbrously draped vertical window, an aperture in a wall, enables the closely guarded interiority of the individual, its inner, private self, to have a social being as it gazes beyond.

In its scopic desire, a culture of the gaze tends to will away this interposing element between antithetical states – to hurry to the clock face, to the illusionistic counterlikeness of the optical game, the print beyond the camera, the longed-for scene beyond the window pane, the goods behind the shopfront, the peach beneath glass. Even, as Foucault understands Bentham's panopticon (in *Discipline and Punish*, 1977), one of the earliest glass and iron structures to be contemplated, the unseen surveillance of the overseer, who need not even be a subject but simply an *idea*, can act directly, in unmediated violence, on the prisoner in his glass cage. And yet the more it is willed away the more this material returns to remind the seer, or user, of its function. Hence glass occasions an intense anxiety of mediation, a scientific, philosophical, aesthetic and civic problem to the Victorians, which is the theme of this work.

Gillian Rose's argument with postmodernity in *The Broken Middle* (1992) and her belief that critique can come about only if thought remains with the broken middle and its difficulties – her book is an attempt to reconstruct an idea of mediation as a challenge to postmodern culture's elision of the middle – are important here. She argues that a peremptory and violent synthesis occurs if the middle is forgotten. One of her examples is Foucault's representation of reason and civic institutions as a compact of monolithic domination. Such totalizing figuring of power as something unremitting and universal is possible only because Foucault has driven power and its objects together in a formal way through simple oppositions, without going through the particular structural processes by which power comes into being, and without considering the specific actuality of power. For Rose, a systematic thinking through of the middle, what is in between one condition and another, actually throws our understanding of the beginning and the end into question. Where to start, a misrecognizing, linear question, becomes a problem. But that is because there is no beginning which is not undone by the unmended middle, no self-evident beginning and terminal points. The middle is always broken, always in equivocation. The result is not a sickening destabilizing of the centre, a state of perpetual flux, as some post-structuralist readings suggest. It is precisely because the centre cannot hold that critique and ethics become grounded. Equivocation moves from one side to the other, making decisions and choices about individual positions. This is very different from being paralysed by radical ambiguity. In this movement thought confronts itself as particular and universal, not merely one or the other. It confronts complexity, not synthesis.

For the nineteenth century, glass can be an invisible membrane, a barrier, or a middle, an equivocating, difficult substance. But the interposing nature of glass, as in the military and political metaphor behind the idea of mediation, as that which intervenes between two enemies, is ever present. In reading glass Rossetti moves between the uniquely blown artifice of Venetian glass and the broken, mass-produced bottle, which forces her to the equivocation of the broken middle. Not everyone did this. Different responses to new forms of mediation, as the infinite possibilities of the prefabricated panel and the lens emerge, create a dialectic of transparency. I am concerned with this, not simply in order to chart experiences of transparency and to make a historical description of the cultural meanings and readings constellating around glass, although this is part of the purpose. 'Phenomenology' as well as 'poetics' are useful terms to describe what I am attempting because they signal a form of analysis which attends to patterns of meaning emerging in cultural experience that may not be theorized by people suffering the full pathos of their lived world. It is a project which tries to work out the dynamic, process and logic of that experience, the cultural forms which emerge from it and the contradictions which belong to a humanly made material at a particular point in history, glass. I am using the word in its classic Hegelian sense. The phenomenology of mind considers the forms and relationships which emerge when consciousness negotiates with itself and the world, a fundamental contradiction which is driven to further contradictions in the effort to remake itself. I am thinking of culture working upon a material: as consciousness works upon the world.

Phenomenology, in the many forms it has taken, has always been attentive to process without being crudely evolutionary: it has retained a complex understanding of the relationship between subject and object rather than dissolving them. Seeing has been one of its preoccupations; that is why mediation has always been important to this form of thought. It has clung on to a form of agency – the way things *work* – while recognizing the power of ideology.

In this abbreviated discussion I want to take this framework of phenomenology and mediation as a provisional model, at least for approaching glass, and to test it out on two forms of transparency, the glasshouse and the optical toy. A poem by Dante Gabriel Rossetti, 'Rose Mary', will become a choric text.

VI

The glass city, invisible utopia. Histories of glass tend to begin at 3000 BC and stop at around 1815, just when glass ceases to become a high art and begins to belong to technology. Just two years after the Battle of Waterloo, John Claudius Loudon was experimenting with curved cast-iron bars which enabled him to construct curvilinear hothouses. He gave the patent for his extraordinary astragals to the firm of W. & D. Bailey of Holborn deliberately, in order that such building forms could be widely available, fully aware that he would make nothing from his invention. He had made his fortune (and was to lose it again), but this abnegation of patent at a point when invention was a saleable property and when share capital could be invested in developing marketable projects is an interesting political move from a man who, elsewhere in his work, demonstrates considerable financial sophistication. Glass architecture springs from the seminal moment of the hothouse, and from Loudon's principled obsessions. The crystal convexities and declivities of his rounded hothouses, their cupolas, rotundas, and, above all, the domed form this practical visionary perfected gleam from the geometrical exactitude of his drawings. The 'junction of different curvatures . . . is to show, that, in regard to form, the strength and tenacity of the iron bar, and the proper choice of shape in the panes of glass, admits of every conceivable variety of glazed surface', he wrote in his *Encyclopedia of Gardening* (1822). The tense forms of the semi-ellipse, the parallelogram with curved roof and ends, but especially the infinitely repeatable combination of iron bar and glazing, presage buildings hundreds of yards, even acres in size. He was the first to read the complex civic meanings of glass. His own house, a 'double detached' residence in Bayswater, was surrounded by a glassed veranda and an *en suite* range of conservatories swept from the house to the extremity of the garden.

Three of Loudon's works were specifically concerned with revolutionary ways of constructing hothouses: but this author of four million words, who habitually dictated his works until two in the morning, returned compulsively to the conservatory whatever he wrote. As a republican and democrat, universal access to independent houses of glass – the lean-to conservatory, the glassed windowsill, the plant cabinet, even the glass-roofed loft to which vines could grow from ground level – signified for him not simply the domestication of technology or the expansion of luxury to artisans and labourers but the activation of desire in urban culture and ownership of self. It opened out, for women, forms of autonomy. The reverse of the retreat to the pacification of the suburb, the conservatory legitimated wanting, permitting workers to want the experiences of enjoyment in excess of subsistence which made for a decent civil society.

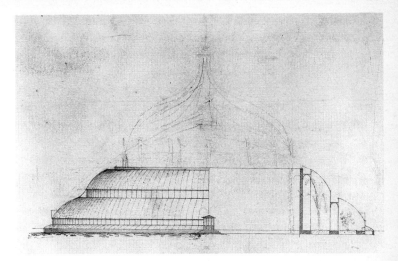

John Claudius Loudon's original sketch of a hothouse plan for the
Birmingham Botanical Gardens, 1830. (Courtesy of Birmingham City
Archive)

It was 'pernicious', he said in *An Encyclopedia of Agriculture* (1825), to expect the poor to be content with 'mere necessities'. Architecture was of necessity a democratic art because everyone could and should have experience of it in daily life.

Both mythos and practical reality, the conservatory was a sign of the virtuosity with which an absolutely artificial environment could be created. Yet this entailed working with two contrary qualities of glass, its property as obstruction, turning back the rays of the sun from its surface, and its property as shield, protecting the micro-atmosphere within the shell of glass. In order to control the growth of the palm or citrus, or to force the grape to fruit at exactly the right moment of prematurity, three months early, it was necessary to be a virtuoso in the management of light. It was necessary to calculate the maximum possible transparent surface capable of being supported by the minimum iron frame: it was necessary, with meticulous violence, to train the sun's rays to fall perpendicularly throughout the day, throughout the year, on a minutely computed roofing slope, cheating the sun of its natural obliquity.

... if 1,000 rays fall upon a surface of glass at an angle of 75°, 299 of these rays are reflected; consequently in little more than an hour after each midday, in spring and autumn, nearly one third of the effect of the sun is lost on all hothouses with parallelogram base and common sloping roofs fronting the south. (*Remarks on the Construction of Hothouses*, 1817)

The dome, ridge-and-furrow roofing, pure crown glass (the most translucent of the glasses) rather than the cheap green panes from which the greenhouse received its name, were some of the resources for controlling light. But as well as a *forcing house* the conservatory was also a *nursery*. Meticulous nurture and protectiveness were required, the exact temperature, the exact degree of moisture.

This double aspect of force and nurture governs the three paradoxes Loudon sees in the glasshouse. It was possible, first, to produce 'artificial climates', exhibiting 'spring and summer in the midst of winter', by means of steam heating (produced by Kewley's 'artificial gardener') and artificial rain (manufactured by the Loddiges firm). And just as steam could be piped and directed under the glass roof, so people could be circulated through a suite of glasshouses as the sequence of their experience was controlled. Secondly, it was also possible to 'bring to perfection the delicious fruits and splendid flowers of the torrid zone in a temperate or cold country'. The vegetation of the exotic and colonial regions could be brought from the periphery to the centre, and the tropical flora of the torrid zones – greenhouse specimens came from China, the Cape, Australia – could displace temperate specimens. On a single page, for instance, Loudon's *The Green-House Companion* (1824) lists specimens from America, China, Mauritius, Canary, the Levant and the Cape. (Such vegetation, indeed, provided the taxonomy of the conservatory. Loudon deliberately moved to a system of classification based on likeness among groups evolved by Jussieu, rather than the Linnaean system.) And colonial produce was not only consumed but exhibited – the exhibition was the scopic pleasure for sale. Finally, the consummate art was to enable fruits and flowers, customarily fruiting and flowering at widely different times in the season, to mature, if not simultaneously, with the narrowest of temporal gaps between their fruition. Loudon gives the most elaborate instructions for achieving, by subtle adjustments of heat and position, the near coincidence of fruiting in peach and grape. The impossible convergences of Christina Rossetti's 'Goblin Market', where fruits are 'All ripe together', were being researched thirty years before she wrote. Again, a conceptual move transfers these conditions to people: Loudon believed that a national education system should teach the children of the rich and the poor side by side.

Three paradoxes, then, a steam-heated micro-atmosphere, the taxonomy of the tropics and the simultaneous maturation of different fruits and flowers, enable the hothouse manager to defeat time and space and give man 'a command over nature': transferred to social and economic relations, they have far-reaching implications which work in contrary directions, towards radical possibility *and* social control. Though it isn't always clear whether the conservatory is a public or a private space, the last refinement of high aristocratic living or a populist experience within reach of the many, a retreat of the rich or a display for the poor, Loudon was always optimistic

– too optimistic – about the conservatory's possibilities. In his earliest work, pointing to the three English acres covered by glass in Imperial Russia, he looks ahead to the point when 'one magnificent range' of conservatories can transform experience, when 'seemingly the most useless of the *debris* of our globe', sand, can create new conditions for living. In 1817, in his work on hothouses, he envisaged an artificially constructed culture:

When subsequent improvements in communicating heat, and in ventilation, shall have rendered the artificial climates produced, equal or superior to those which they imitate, then will such an appendage to a family seat be not less useful in a medical point of view, than elegant and luxurious . . . Perhaps the time may arrive when such artificial climates will not only be stocked with appropriate birds, fishes, and harmless animals, but with examples of human species from the different countries imitated, habited in their peculiar costumes, and who may serve as gardeners or curators . . .

From emancipation to the human species as display. This looks forward to Loudon's later vision of a public utility in the form of acres of glass-covered, steam-heated landscape, farmland and parkland combined. This in turn looks forward, carrying all its contradictory meanings of exploitation and emancipation along too, to the vast conservatory and display space for commodity that was the Crystal Palace of 1851. The proposals for crystal landscaping of the city's routes were ceaseless. In 1845 Frederick Gye proposed a seventy-foot-high glass arcade from the Bank of England to Trafalgar Square, with shops, meeting rooms, libraries, cafés and a flower market. In 1855 William Mosely proposed a 'Crystal Way' from Cheapside to Piccadilly Circus, and in the same year Joseph Paxton supplemented his Crystal Palace model by presenting plans for a Great Victorian Way to a parliamentary committee on relieving congestion in London. It was a seventy-two-foot-wide crystal 'Girdle', a circular promenade, over ten miles long, encompassing central London with access to rail termini, and flanked by eight raised railway lines.

The new language of the glass landscape was not wholly about an unreal city, it was about a crisis in how to conceptualize the city. It was about financing with share capital the mediation of traffic, of people and of goods, and how to think about what the Select Committee on Metropolitan Communications in 1855 called 'those great tides of passenger traffic which . . . are flowing through our streets'. People as anonymous flow, to be conducted, like the effluvia of graveyard and sewer, or like the polluted water of the Thames, into new routes and pathways, were deemed to be the problem to be solved. It is significant that in 1850 Paxton collaborated with Chadwick, the foremost campaigner for sanitary reform in cities, to design new London cemeteries which incorporated a glass-domed church and glass-covered ways. Ever since Fourier considered glass arcades essential for citizenship of his utopia early in the century, it seems that the ideal of planning was two-sided, to enable people to move and to move them along. 'The protective connecting routes are one of a thousand amenities kept in reserve for the new social order.' In both Gye's and Paxton's crystal boulevards the residuum of the conservatory exists as plans for a flower market in these glass domains, surely an overdetermined detail. Cut flowers, the ultimate city commodity, luxury product of the artificial climate of the greenhouse, are a memorial to its possibilities as well as signifying the culture of commodity display for consumers. Loudon's artificial climate now provided a means of transcendentalizing the urban flow, creating a ceaseless hydraulic circulation, like the movement of steam in his hothouses, separating out people from the immediate conditions of their experience and from their waste products, which circulated in the ground beneath. It was implicitly conceived as protection, protection from the dark, unmapped zones of barbarism where the poor and criminal underclass lived in unprecedentedly squalid conditions. These were metaphorized as the savagery of the colonial other, the black, unregenerate territory of Africa in particular. Strangely, those areas of amnesia encouraged by the platonic glass pathways (with their attempt to make everything transparent, even death), the stench of waste and the intrusion of the colonial subject actually reproduce a transcendentalized form of the problem in an attempt to eliminate it – the circulation of waste, the artificial climate in which the exotic can breed and be exhibited.

Is the model of forcing house sustained in the glass channel as peremptorily and violently unmediated as Bentham's Panopticon? Or does it display an anxiety about technologies of social movement and control? A poetics of glass will have to explore this further. Meanwhile, it is interesting to know, as his wife's memoir of him tells us, that Loudon was present at the lecture at which Southwood Smith dissected and mummified Bentham's corpse.

In declaring in a remarkable retrospective lecture in 1851 that the Crystal Palace was an 'exhalation' (he was right, of course; it was made of breath held in frozen glass) and affirming that the exhibition overcame the conditions of space and time, William Whewell, Master of Trinity College, Cambridge, was only following through the logic of the greenhouse into the logic of capital. As the exhibition annihilates time by bringing together the products of nations at very different stages of their historical development, heralding standardization, so it annihilates space. Succession can be reduced to simultaneity if, by thinking of ourselves travelling a little faster than light, we 'overtake successively the visual images of all successive events', perceiving them in 'inverted order'.

We might thus see now what is passing around us, and the next minute, by rushing to the borders of the solar system, where the images are still travelling outwards, see the first inhabitant of this island . . .

Without this new order of simultaneity, the individual subject's power to control and organize the world, an experience offered even if illusorily by traditional ways of knowing, is diminished, because he or she cannot possibly encompass the totality of things. Whewell invokes the lens of the camera and the telescope to envisage the solar system of new images of mediation which the hothouse of the Crystal Palace has generated. I turn to the lens and its meanings, but not before remembering that texts disclose the invisible presence of an invisible layer of glass in innumerable ways. The conservatory is the scene of revelation or reversal, as in Meredith's *One of Our Conquerors* (1891), or in Christina Rossetti's 'At Home', where the excluded, disembodied ghost hovers in the place of botanical nurture, seeing its family 'Feasting beneath green orange boughs',

The glass fountain at the Great Exhibition of 1851.

conversing of time with the hubris of those for whom time can
be manipulated. But it is the absence of the 'diaphane' which tells
us most about mediation, particularly in the novel. The artificial
climate of culture and its hidden 'colonial' presence, at home and
abroad, the taxonomy of display and the subject's impossible syn-
chronicity of experience, the ethics of the gaze, all these are narrative worries which call out new resources of narrative.

VII

For how much mankind depends on this elegant material produced from seemingly the most useless of the *debris* of our
globe . . .

VIII

The year Loudon published his first treatise on hothouses, in 1817,
Sir David Brewster wrote in his *Treatise on the Kaleidoscope* that
200,000 kaleidoscopes were sold within three months after the first
demonstration of this instrument for playing with colour and the
redoubled image in reflection. The rage for optical toys of all kinds,
deriving from the lens and the screen, reflection, refraction, projec-
tion and anamorphosis (or distortion), generated innumerable scopic
games. The magic lantern, the solar microscope, the transparency,
the stereoscope, the pseudoscope (the negative form of the stereoscope where images were concave rather than convex), perspective
games, polyoptic pictures, mirror metamorphoses, concertina views
were forms relating to the small-scale peepshow which began with
the camera obscura and continued with the microscope. (Barbara
Stafford, in her magnificent *Body Criticism* (1991), has described
some of these gadgets.) The panorama, in which you moved before
a massive pictorial image, often lit and used as a screen surrounding
you in a circle, and the diorama, in which images moved in front
of you, were large-scale spectacles which organized the whole
body as well as isolating and intensifying the optical pleasure of the
hyperactive eye. Among these large-scale scopic gadgets must be

included the optical illusions available in street and pleasure garden.

What did all this mean? First, it meant repetition, the experience of again and again, the incessant visual surprise whose shock was actually in the pleasure of its repeatability. This appears to betoken *control* of its experience by the gazing subject because the *trompe-l'oeil* can be perpetually manipulated. Like the child's game Freud designated as *fort da*, where a cotton reel is thrown away and retrieved to control the mother's absence, the eye plays with disappearance and retrieval of images. Second, it meant a sensory experience without a sensory, tactile image, an image that was more spectral the more reproducible it became, so that the gazing subject played a risky game with materiality and phantasmagoria – not just seeing magic but exploring the illusion of working magic with the world through simulacra and reflection. A parody of Herschel by R. A. Locke in 1836 describes how a wonder telescope situated at the Cape throws on to a screen the magnified images of moon cattle grazing in lunar fields. The creatures could be waved at and touched, but continued their grazing undisturbed. Thirdly, it meant mystery. Where was it? Here is the excitement of losing control. Gye produced spectacles of baskets of fruit which receded as you approached them. The figure of a man would appear on a dead wall. Micro-images could be enlarged, displaced, reflected, refracted. Do they come from behind, in front of, the viewer?

The pleasure of deception itself, where the lens can produce deformation, where the atomized detail of the subvisible world on which the microscope relied can dissolve certainty, confusing the distinction between convex and concave, could be a kind of epistemological masochism. Possibly the intense allure of these gadgets was their capacity to suggest experiments with different subject positions, control, displacement, obliteration, power, centrality, powerlessness. Arguably, they redeemed the anonymity of the glass forcing house through play.

But it is well known, that the reflected image of any object, when placed before another mirror, has an image of itself formed behind this mirror, in the very same manner as if it were a new object.

Sir David Brewster's observation, in his account of the kaleidoscope, presages the infinite renewal of the world as well as its infinitely phantasmic nature.

Ruskin did not think so. *Modern Painters* was partly written in rage against a *trompe-l'oeil* culture pathologically deceived by the distortions of anamorphosis and the reflection conceived as a 'new object'. He hated particularly the private, possessive, monocular vision of the microscope and anything that derived from it, penetrating the hidden world, erotically unveiling the subvisible, and wanted to restore the full democracy of the eye. What it is to *see*, as well as to see *what* you see is at issue. For in reality optical gadgets *mastered* the subject, not it them. In *Stones of Venice* (1853) Ruskin used the anamorphic image of an optical toy to describe the 'terrible', oppressed grotesque consciousness which was so benumbed by sensational experience that it was incapable of critique or creativity. With 'strange distortions and discrepancies', perception plays upon the broken surface of a mirror, 'all the passions of the heart breathing upon it in cross ripples'. Ruskin analysed the consumingly desiring consciousness, obsessed with limit and death, because it knows what it sees as representation is distorted, but is entrapped in ever more distorted experiences the more it tries to correct them. This is a culture of violence, which is violent because it has lost forms of mediation which can escape from ideological control. The cross-ripples of the fractured mirror bend and distort light, and this refraction becomes for him the very image of ideology, which catches us in a world of bent images.

Sir John Herschel had indeed seen the earth's atmosphere as a transparent film, like glass, bending light. He pointed out that a prerequisite of astronomy was to understand that the rays of light you saw were bent so that the astral object appeared to the side of the place it actually occupied, and the 'actual' rays of light from the object fell behind the viewing subject, who 'actually' saw other rays of light. Almost as important as this displacement of the viewing subject is the fact of the earth's motion. This means that anything like the idea of the earth as a fixed point has to be given up. It is important, he says in his work on the telescope, in 1861, to relinquish the notion of the earth as a foundation. Our motion in relation to that of another object is what has to be calibrated. With this we move to a world of continuous relational adjustments. Speaking of the kaleidoscope, Brewster made the same point. Looking is always triangulated: there is the viewing subject, the viewed object and the instrument you view it with – and all can be in motion. For Ruskin the drawing subject is the patiently stationary subject, but the lens as described by Herschel and Brewster creates a world of vertiginous motion and constant adjustment, the risky moment when you get the image 'right'. Benjamin thought of the accelerated images of the diorama as the equivalent of a street of shops moving at speed before the eyes of the dazzled consumer: we can add to this the fundamental instability of the universe (of capital also?) as the viewing subject attempts to come to terms with it. Did optical gadgets probe this deep instability of the scopic unconscious?

IX

The choric poem 'Rose Mary', by Dante Gabriel Rossetti, comments on the transcendence of time and space which the hothouse proposes, the annihilation of the subject as agent, and on the addictive, repetitive recourse to optical toys and perceptual games as a form of knowledge. The great 'error', if it is an error, thought through in the poem is to mistake signification for sign, simulacra for the same. What status do they have? Embodying in its title the contradictory states of Experience or sexuality and Innocence or virginity (Rose and Mary), and, through their linkage, Rosemary, memory or representation – and memory is almost by definition an *unstable* representation – it explores the knowledge which comes from looking into glass. It is in three parts, but the dialectic of eros and purity does not produce a synthesis.

The whole poem is organized round accelerated images seen in a glass. A mother, anxious for the safety of a daughter's lover, and knowing the daughter's divining gift, looks through her daughter,

by proxy, into the transparent beryl stone, a magical stone stolen
from Moslem hands long ago.

> With shuddering light 'twas stirred and strewn
> Like the cloud-nest of the wading moon:
> Freaked it was as the bubble's ball,
> Rainbow-hued through a misty pall
> Like the middle light of the waterfall.

They see the moving images of the road by which the lover will
travel: 'the road shifts ever'. 'Swift and swifter the waste runs by /
And nought I see but the heath and the sky.' In desperation the
mother urges her on: 'One shadow comes but once to the glass.'
Doubled images and perspective tricks occur as the girl compulsively
repeats the act of looking. Different montage effects of the same
image flicker into vision.

> The glen lies deep and the ridge stands tall
> What's great below is above seen small,
> And the hill-side is the valley-wall.

This compares with:

> Again I stand where the roads divide;
> But now all's near on the steep hillside,
> And a thread far down is the river tide.

The ambush is seen in the glass, but not where the lover is riding.
Yet the road *has* 'shifted' (deceived) the lookers: the lover's dead
body is received at the castle, and the mother sees from a letter
in his clothes that he was actually on his way to meet another
woman. The daughter smashes the beryl glass for misleading her,
but does not know of her lover's unfaithfulness. But it is the lover
and her own feelings, not the glass itself, which have deceived her.
Moving from the specular stone into the day the 'Open light most
blinded her.' The outside world returns the same images as the
stone. *Its* images determine how she sees in the day. At the climax of the poem the mother runs down the winding stair of the
castle, whose loopholes show fragmented images of valley, wood

and stream, the 'real' images represented in the glass, like 'torn threads of a broken dream'. The loopholes, like the moving images seen through the slits of the diorama, implicitly ask whether the subject's movement or the image's movement, or both, create meaning.

Rossetti is not simply concerned with a replication of the visual conditions of the optical toy, however skilfully he realizes them. The poem addresses several problems. First, to smash the stone is to misunderstand what has occurred. The lover himself and the daughter, who has consummated her love with him, create the conditions of deception. The written, spoken and visual representation can all 'shift' ever and cannot represent or match the hidden conditions of the actual. Secondly, there is an ethics of seeing. Why is the mother so avidly desirous of the vicarious experience of seeing, looking by means of somebody else's sight? At the conclusion of her 'vision' the daughter is exhausted, and becomes like a stone, the condition of the labourer whom Ruskin describes as the victim of the division of labour. We use the labourer's body and breath, seeing through it vicariously. Here, transferred to the relations between two generations in a family, the implications are sinister. Next, there is the problem of the visible. 'Fear no trap that you cannot see', says the mother. If you are unable to see a trap, don't go out of your way to look for it, is one reading: another is that the world is full of unseen traps, whatever the evidence in the glass. And this evidence is itself unstable, speaking the truth by 'contraries'. That is, 'truth' will reverse the evidence of cognitive experience: like Brewster's phantasmically 'new' image, it is a reversed or 'double' image. Thus there is no question of seeing *through* the image or representation to reality. Where the ambush actually was, there was a thin patch of mist, impossible for the soldiers to hide behind, they think, and ignore it. But they have not so much misrecognized the nature of the mist as concealing the ambush; they have failed to see its *reality* as blind spot, for the mist *is* the reality. That is what the mist reveals rather than conceals. Its truth is its nature as simulacrum. Rossetti here comes close to reading the blind spot as ideology, and desire becomes an aspect of ideology.

The daughter scans the moving road not once but twice. It is as if this repetition creates two concurrent times and spaces, forcing both the past and the future into the present. The act of looking in the present is parallel with the penetration of a future time and space as the girl tries to divine, and a past time and space as she checks on her divination. But such is the untethering of time and space in these repeated attempts to see into another dimension that they become, simply, different times and spaces, coexisting. How far this experience, which is actually one of both fragmentation and simultaneity, enables the viewer to be in control of the images she sees is thus thrown into question. Desire orders images, but how far do images control desire? The frantic, ever-accelerating desperation seems to be trying to keep pace with the production of images – and their destruction. In this poem Rossetti sees deep into the perplexities of glass in his culture, a culture which is ours also.

'Wireless': Popular Physics, Radio and Modernism

GILLIAN BEER

'The isle is full of noises': since the coming of radio it has become impossible for us quite to feel the charge of energy and excess in that line. We take it for granted that the isle *is* full of noises: that bodies and voices can readily be detached, that heard instruments are the vagaries of the air, that our neighbour's choice of recorded music will become part of the deep beat of the building. Radio, of course, is not now the only source of such vagrant sound. Whistling is rare nowadays but television, CD-players, tapes, augment the occa-sional live street musician. Only in the case of such musicians are body and sound conjoined. This causes us no difficulty of apprehen-sion. Earlier, such disjunction required magic, an isle full of disem-bodied sounds become the quintessence of the magical. Now, the one idea that still surprises is that the soundwaves do not cease when we turn off the radio; they continue to course through the air, inaudible but not cancelled. We switch in and out of a constant stream of potential sound to which our radio gives us access.

This experience of sound disconnected from its source (a source distant in space and perhaps time, always readily available) is quite novel in the history of the world, the history of sensibility. Yet it goes, in the main, curiously unremarked. Perhaps this is because of its intimacy: sound uncoupled from other sources rapidly enters and becomes part of taken-for-granted inner experience. In Britain, at least, rather little work has been undertaken on radio, while film has been given a privileged place in social and media history. (One exception is Paddy Scannell and David Cardiff, *A Social History of British Broadcasting* (1991) but so far only volume one, 1922–1939, has appeared.) Asa Briggs's magisterial five-volume study of the BBC concentrates on the public, even authoritarian politics of the institution; but radio raises other questions too. A broadcast is, intimately and fleetingly, part of the mind of the listener, leaving an unwritten trace and the subject of unvoiced dialogue. So when, for example, we are thinking about the interconnections between writers after 1928 (when radio broadcasts became easy to tune into without interference) we cannot know extensively who tuned into whom or how hard they listened. Arthur Eddington and James Jeans, Virginia Woolf and Vita Sackville-West, E. M. Forster and Bertrand Russell may each have heard any of the others on the airwaves. Some among this skein we know for certain did so. Who has read whom is no longer (if it ever was) a sufficient question: the relations of the written and the oral are rearranged through the often pseudo-orality of radio, invisibly scripted, purportedly dialogic.

Radio produced a new idea of the public, one far more intermixed, promiscuous and democratic than the book could cater for. The fiction of one to one in radio address is very different from the social sweep of film in its public darkened place of communality, the cinema. Music on radio was one of the immediate great gains in access, though the live conditions of its production may take us aback now. Sylvia Townsend Warner records in her diary for 12 May 1928 that she went to a (private) live performance in Broadcasting House of Stravinsky's *Oedipus Rex*, with Stravinsky himself conducting:

A great deal of the choral writing was almost pure Taverner, the construction that of an early passion. It is really impressive music, it sounds old and cold, as a chilly shadow that has never lifted from man's mind. Two incidents disturbed our listening. First a man stole in and with portentous clattering of keys unlocked a cabinet, taking out a soda-water syphon and a decanter of whiskey; then my cigarette case of synthetic jade burst into flames with an overpowering stench of camphor.

The Diaries of Sylvia Townsend Warner, 1994

Radio speeded up reception: ears take time to learn new sound worlds but from the start the BBC included contemporary music in its programmes.

On radio, material for thought that had earlier pointed towards a particular audience with particular expertise now had to be re-thought as 'general knowledge'. The idea of the 'general audience' could produce a new form of bland authoritarianism in which the speaker and programme maker preselects what the listener is sup-posed to be able to grasp. But in the first years of the BBC it also produced an energetic attempt to address the listener as an equal in intelligence if not in technical information. Of course, these two positions sometimes overlapped.

The new audience was both anonymous and intimate, not likely to be listening in a crowd, going about their own household pursuits as if in family conversation, or sitting in the armchair ready for a private lesson, yet (for the broadcaster) almost truculently invisible, unknown, perhaps derisive or at the least inattentive. What held them together was the English language and a newly forming and changing British identity, class-bound as ever, yet opening up what some politicians and commentators saw as a dangerous 'radiocracy', making available a range of ideas to people who then *used* them, without the traditional licence of higher education.

How does this chime in with the abstruse, sometimes coterie, conditions of modernism, a loose assemblage of writers such as Woolf and Joyce, H. D. and T. S. Eliot, whose self-descriptions often set them on the opposite pole from science (or indeed democracy)?

In his recent book-essay *We Have Never Been Modern* (1993), the influential philosopher of science Bruno Latour argues that what he calls 'the delicate networks traced by Ariadne's little hand' have been rendered invisible by the forceful disjoining of various intellec-tual sets. The sets are: those concerned with things-in-themselves or nature; those concerned with power; and those concerned with texts and discourse. Latour's insistence on interlacing is attractive. But can we accept the sanguine implication that once such intricate webs are established a more benign condition is likely to prevail? It is never clear in his argument whether the ravelling of networks – the rendering them visible – is to be simply a satisfaction to the scholar or to produce a more harmonious social order. The sugges-tion is that Ariadne's knitting will produce a seamless garment, or the spider Arachne's high-tensile web. But it may equally produce garble, gabble, interference on the airwaves. The scholar's seamless story will be, in the nature of narrative, a retrospect, thickened and clarified, with power over the past, not necessarily issuing into the present. Moreover, what Latour describes as delicate networks of connection are delicate – and sturdy – epistemological webs. Webs are traps as well as support groups, or linen.

In the late 1920s and 1930s, at the height of scientific and liter-ary modernism and during the anxious years that climaxed in the Second World War, attempts *were* made by scientists, philosophers, politicians, writers, radio listeners, to gain a purchase on communal experience by putting pressure on each formation. In that interac-tive community, nature, power and discourse do not seem far apart. That example is the one I concentrate on here, and on the interacti-vating potential of a then fresh technology, radio, and its scientific source, physics.

When radio first became generally available it was more often called 'wireless' with the emphasis on its freedom and its lack: look, no wires! Sound cascades through the wave systems of the universe, uncoupled from any visible medium of transmission. Instead of sub-stance and sequence, quantum mechanics concerns itself with waves and scale, as Professor Bragg put it in a radio talk then published in *The Listener* in June 1929:

we can range from the vast wireless wave which is like an ocean roller a mile from crest to crest, through the ripples of heat and the minute ripples of light which are one fifty-thousandth of an inch apart, right up to X-ray waves, which are ten thousand times smaller still. The end is not there, for the cosmic rays recently discovered by Millikan are far finer still.

Strikingly, physicists were among those most willing to take the risks on radio of communicating with an unknown audience and giving entry to unstable and disturbing understandings – not of distant worlds only but of humdrum tables. Arthur Eddington, indeed, in a double reflexivity, employed radio, itself the medium which was the outcome of physics and technology from Hertz, Marconi and Clerk Maxwell, as a principal channel of communication for the explanation of physics. Many of his books were the outcome of series of radio talks. And he also used the technical features of broadcasting as a metaphor to explain the implications of those theories. In *Science and the Unseen World* (1929) he used the idea of broadcasting stations to bring out the incommensurability of sign, signal, utterance, place.

The mind as a central receiving station reads the dots and dashes of incoming nerve-signals. By frequent repetition of their call-signals the various transmitting stations of the outside world become familiar. We begin to feel quite a homely acquaintance with 1LO and 5XX. But a broadcasting station is not like its call-signal; there is no commensurability in their nature. So too the chairs and tables around us which broadcast to us incessantly those signals which affect our sight and touch.

The everyday substantial world is no longer solid but itself transmissive. Wireless makes intermittently manifest the invisible traffic passing through us and communicating by our means. As Eddington puts it in *The Nature of the Physical World* (1928):

In the scientific world the conception of substance is wholly lacking, and that which most nearly replaces it, viz. electric charge, is not exalted as star-performer above the other entities of physics. For this reason the scientific world often shocks us by its appearance of unreality. It offers nothing to satisfy our demand for the concrete. How should it when we cannot formulate that demand?

In such a newly imagined world wireless becomes more than a metaphor for the almost ungraspable actuality of the universe. It is the technical realization of new scientific imaginings, a realization that itself materializes their improbable possible worlds. The geneticist J. B. Haldane, in his 1927 essay on 'Possible Worlds', remarks that 'Einstein has left common-sense space in a badly damaged condition.'

A century ago physicists began to give up the corpuscular theory of light, which had satisfied them for two thousand years, in favour of a wave theory. Among the practical consequences flowing from this theory were wireless telegraphy and telephony.

Out of that theoretical shift have come the practical consequences – telegraph, telephone, wireless; those practical consequences themselves confirm a conception of the universe newly magical. So, in a handbook of 1928, *Everyman's Wireless*, Ernest Robinson combines practical instruction with cosmic explication. Radio is a manifestation, close at hand, of the nature of life itself:

everything we know about the physical universe is based on the association of one or more electrons, or negative electric charges, with a positive charge of electricity. Rocks, air, water, trees, and flowers, the fish that swim, the birds that fly, our own bodies, the food we eat, the cars and trains and trams we ride in, are all, when we get down to the very bottom, simply made up of charges of electricity grouped together in countless different ways.

As Orlando reaches this same year of 1928, in Virginia Woolf's fantasy biography, s/he thinks how technology has remade the world as magic:

So she stood in the ground-floor department of Messrs. Marshall & Snelgrove . . . Then she got into the lift, for the good reason that the door stood open; and was shot smoothly upwards. The very fabric of life now, she thought as she rose, is magic. In the eighteenth century, we knew how everything was done; but here I rise through the air; I listen to voices in America; I see men flying – but how it's done, I can't even begin to wonder. So my belief in magic returns.

Really new ideas are extraordinarily difficult to grasp; the mystery is that in fifty years most of the same ideas have become easy, taken for granted, taught in schools. This phenomenon of the abstruse becoming easy is, in its turn, taken for granted. But the process requires extremely complicated explanation and indeed could be said to be the underpinning enigma of cultural studies. True, as Haldane points out in 'Possible Worlds', what is 'of the greatest importance for physicists today' may be 'of the greatest practical importance to our grandchildren'. Application elucidates – or does it?

Wireless and telephone are not part of an earlier world of 'common sense' – indeed they go against all its criteria: instead they produce disembodied voices, action at a distance, access to the tumult always at work in our silences. We switch on, tune in and hear the airwaves. They do not cease when we switch off.

The sly magic of amateur radio is brilliantly evoked and skewed in an earlier story by Kipling, 'Wireless' (1903), which opens:

'It's a funny thing, this Marconi business, isn't it?' said Mr Shaynor, coughing heavily. 'Nothing seems to make any difference, by what they tell me – storms, hills, or anything.'

The setting is an Edwardian chemist's shop where an attempt is to be made through Morse code to signal and listen in to ships at sea off the coast in Dorset. The manner of the tale is one of lower-middle-class realism with a considerable dose of hyperbolic exposition:

'You've got a charged wire –'
    'Charged with Hertzian waves which vibrate, say, two hundred and thirty million times a second.' Mr Cashell snaked his forefinger rapidly through the air.
    'All right – a charged wire at Poole, giving out these waves into space. Then this wire of yours sticking out into space – on the roof of the house – in some mysterious way gets charged with those waves from Poole –'
    'Or anywhere – it only happens to be Poole tonight.'

The intended experiment is not a complete success, but what seems a much stranger link also occurs. One question the story proposes is: why does this other link seem stranger to us?

'But what is it?' I asked. 'Electricity is out of my beat altogether.'
    'Ah, if you knew that you'd know something nobody knows. It's just It – what we call Electricity, but the magic – the manifestations – the Hertzian waves – are all revealed by this. The coherer, we call it.'

That technical word 'coherer' might be the title of Kipling's tale, which menaces the reader with the suggestion that the universe hangs together in ways we cannot control or understand. Or if we apprehend, apprehension comes both too early and too late. By the end of the story the coughing Mr Shaynor has become evidence that radio waves move secretively and effectively through time as much as space. For the reader, the story also raises the question of how it is that, in the act of reading, we take for granted moving freely back and forth through time, yet are disconcerted when it occurs as event within the fiction. Evocations and prolepses are melded into the early part of the narrative, but so dispersed that they are at first inaudible: the druggist's store, St Agnes' church, the bitter cold, 'the three superb glass jars red, green, and blue', 'those bubbles – like a string of pearls winking at you', Mr Shaynor's girlfriend Fanny Brand, his terrible cough for which he takes 'cayenne-pepper jujubes and menthol lozenges'. We see it all, too late. We may pick up the mental vibrations early but they delay functioning as intertextuality. Keats is absent. We lack the coherer to produce 'the manifestations – the Hertzian waves' until Mr Shaynor has become our medium.

Kipling evokes the amateur order, muddle and obsessionality of the chemist's shop. Echoes of the alchemical are everywhere, and the force of electricity is a female Power:

He pressed a key in semi-darkness, and with a rending crackle there leaped between two brass knobs a spark, streams of sparks, and sparks again.

'Grand, isn't it? *That's* the Power – our unknown Power – kicking and fighting to be loose,' said young Mr Cashell. 'There she goes – kick-kick-kick into space.'

As this happens, poor young Mr Shaynor, consumptive and obsessed with the shopgirl Fanny Brand, waits for Poole to 'answer with L.L.L.' and (his eyes fixed on an advertisement of a simpering young woman 'bathed in the light from the red jar') begins, trance-like, to write. The first line he writes is

And threw warm gules on Madeleine's young breast.

Mr Cashell in the inner office remarks that 'There's something coming through from somewhere; but it isn't Poole.' Indeed not: it is the druggist John Keats in the throes of composition, now become (or possessing) Mr Shaynor, who has never read a word of him. Shaynor scratches down and gasps through drafts of the poems, experiencing 'St Agnes' Eve' in hectic revision. He is in total identification with the unknown Keats – an identification overdetermined, the first-person narrator suggests somewhat self-mockingly, by

the Hertzian wave of tuberculosis, *plus* Fanny Brand and the professional status which, in conjunction with the main-stream of sub-conscious thought common to all mankind, has thrown up temporarily an induced Keats.

Fantasy, of course; nonsense, maybe: but the tale catches the reader up short. What authenticates? Why do we, knowingly, reject this as fiction? Do we, do *I*, personally have the scientific knowledge to do so: have we given over our faith from religion to science? Max Born in *Einstein's Theory of Relativity* (1924) writes that the world of the physicist is one of 'inaudible tones, invisible light, imperceptible heat' – 'Now, Einstein's discovery is that this space and this time are still entirely embedded in the ego, and that the world-picture of natural science becomes more beautiful and grander if these fundamental conceptions are subjected to relativization.'

On what grounds can electricity and waves moving in every direction be credited by the non-scientist – who takes it as axiomatic that the voice of experience can sound across time only in the written word? We have no applications of physics that give entry to the past, we may reply. Yet Kipling's tale heralds the coming of the moment when that inhibition is no longer quite valid. Voices on the airwaves can now survive to speak across time, though in such strange accents and at such odd pitch. But they cannot hear us. Time is not (yet) interactive.

In a scene between black comedy and desolation at the end of Kipling's story the men listen together to 'a couple of men-o'-war working Marconi signals off the Isle of Wight' and the narrator gazes on Keats's words received and made *now* as in a séance, or in science. The ships 'eavesdrop across half South England' on each other:

Again the Morse sprang to life.

'That's one of 'em complaining now. Listen: "*Disheartening – most disheartening.*" It's quite pathetic. Have you ever seen a spiritualist séance? It reminds me of that sometimes – odds and ends of messages coming out of nowhere – a word here and there – no good at all.'

That slippage – séance, science, signs – touches a nerve in the public response to wireless and to physics in the late 1920s and 1930s. Wireless is domesticated magic only if you desist from tinkering with your set. Otherwise it is garble, gabble, interference.

In his 1928 handbook Ernest Robinson explains:

Waves start off from the aerial through the ether. The amplified speech or music waves, which are at audio-frequency, are applied to this radio-frequency carrier-wave and cause it to spread out and vibrate within itself at audio-frequency. The carrier-wave radiates out in all directions at the speed of light, taking the speech or music modulations with it.

Science, and particularly physics, by the end of the 1920s seemed to posit a universe of what Gerald Heard, the immensely influential 1930s popularizer of science, called in *Science in the Making* (1935) 'cosmic nonsense' – a nonsense which authoritatively undermined all previous attempts to make sense.

Haldane argues that modern physics has gone further:

Heisenberg and Born in Germany, and Dirac in Cambridge, are busily clearing away these vestiges of common sense. In the world of their imagining even the ordinary rules of mathematics no longer hold good. The attempt to build up a world-view from the end which common sense regards as wrong, is, at any rate, being made, and with very fair success.

Such a view leads to systems in which 'the real elements in the external world [are] the secondary qualities of colour, tone, and so forth, rather than the primary qualities of the materialist's world'. Haldane's comments here are not unusual for the time. That is revealing. In a long review of Eddington's *The Nature of the Physical World* in *The Listener* in April 1929, E. A. Milne concludes by pointing out that there is no *hors texte* to nature. The observer cannot stand outside the universe:

But we cannot view the totality of reality externally. We are, ourselves, part of the world we are studying, and we cannot dissociate even the metrical part of our science from our minds. The mathematician [and here he probably has in mind D'Arcy Thomson] describes the ripples of the pond in terms of his own symbols; the poet describes them in different symbols. The mathematician is not alone the freeman of the universe.

Almost all these commentaries emphasize the move away from 'common sense'. As Clifford Geertz points out in *Local Knowledge*, common sense is a term used to imply intuition *beyond* culture and yet is a phenomenon peculiarly closely framed *by* culture. Common sense gains its intuitional intensity because it is formed from *unexamined* assumptions, deeply shared by a community in a historical moment. It is a term of self-congratulation too, drawing to itself words like 'sturdy'. In that, also, it was out of key with the temper of physics then. Yet it may be that to rebuff common sense itself becomes a 'common-sense' move of the time, a cliché which can produce a counterfactual, sustaining, question-closing, communality for literary modernism, under the wing of contemporary physics. Scepticism and credulity become one.

Uncertainty and a sense of being part of what is conditional in your own experiment haunt the writing of Eddington, Bohr, even Einstein. Gerald Heard, always keenly tuned to the mood of the moment, remarks in his article on current 'Research and Discovery' (*The Listener*, May 1930) that 'though a 19th century poet did cry "Glory to Man in the Highest, for Man is the Master of Things", I do not think a 20th century scientist would echo that cry'. The fascination with what lies beyond common sense opened up by physicists in this period shows in the holy books of allusion repeatedly used at the time by scientists and philosophers alike. Hooker's epigraph runs:

> And now, if e'er by chance I put
> My fingers into glue,
> Or madly squeeze a right-hand foot
> Into a left-hand shoe . . .
> I weep . . .

Alice in Wonderland and *Alice Through the Looking Glass*, mathematicians' puzzles, are used to secure the unyoked space beyond common sense back into a domestic zone. Eddington cites them, particularly *Looking Glass*, frequently throughout his work. Alice studies 'uglification'. The Alice books are used as what I'll call 'nurs-erification' – a teasing acceptance of new ideas based concessively on their absurdity, since they cannot be wished away.

Human beings are adept at living in multiple – and conflicting – epistemologies. We could not survive otherwise. The physicist who denies substance must, as Eddington observes, jump out of the way of the approaching car. In ordinary experience we are practised at interdisciplinarity, occupying multiple subject positions daily, gendered, classed – and positioned multiply as tradespeople, lovers perhaps, shoppers, parents, subjects or citizens in the political order where we live and outside which we cannot be. We share common systems of information as well as local and familial ones.

In retrospect we group our modernists as particular literary writers, but:

Your modernist, then, describes nature in terms of a four-dimensional conception, which is neither space nor time, but a complete union of both ... Whether the space–time continuum is regarded as a very efficient diagram or whether it is regarded as an accurate model of reality is perhaps a matter of choice rather than of argument.

Not a description of Joyce or of Woolf or indeed of J. B. Priestley, but (as the reference to diagram and model perhaps reveals) an account of what the writer calls those 'ultra-modernists' 'Jeans, Eddington, Einstein' (the writer in question being C. W. R. Hooker, in 1934, in *What is the Fourth Dimension? Reflections inspired by a pair of gloves*).

At the period that we identify with literary modernism the term modernism had many applications, quite as much in science as in literature. Indeed, literary modernism is a largely retrospective label: at the time writers and readers were by no means clear as to who was or was not a modernist. The term was used also, aggressively, diffidently, in other domains of thought. *The Listener*, for example, in its first year of publication, 1929, advertises works on theological modernism. Evolutionists combined genetics and natural selection from the 1920s on to reach the 'modern synthesis'. So the usual literary application of the term modernism solely to the arts obscures the energies playing back and forth with destabilizing scientific theories at the time.

The writers whom we, in retrospect, distinguish as modernists were at large in a society where popular scientific books such as those by Eddington and Jeans were major bestsellers and where the advent of radio brought conversation about such topics into the house. If you look at issues of *The Nation* (the *New Statesman and Nation* in the later 1920s), you will find Leonard Woolf reviewing the scientists James Jeans and Gerald Heard. And in her letters to Ethel Smyth you will find Woolf reading Jeans and trying to imagine space bending backwards just before she begins writing Bernard's culminating soliloquy in *The Waves* (1931). In the first four years of *The Listener*, from 1929, literary and scientific talks mingle and even overlap. From the early 1920s on Einstein is to be found persistently alluded to in newspapers such as *The Times*.

Whether or not particular modernist writers *believed* themselves to be antagonistic to, or ignorant of, scientific developments they were nevertheless imbued with them, often in a slack and popular way. And that way, chagrining though it may be to scientists themselves, is most often the path by which ideas open out their implications in culture: 'All was fading, waves and particles, there could be no things but nameless things, no names but thingless names' (Beckett, *Molloy*). Or, as Jeans more reassuringly put it in *The Mysterious Universe*, 'we need find no mystery in the nature of the rolling contact of our consciousness with the empty soap-bubble we call space–time, for it reduces merely to a contact between mind and a creation of mind – like the reading of a book, or listening to music.'

On the radio Harold Nicolson was the foremost interpreter of literary modernism for a general audience at this time, just as Gerald Heard was for scientific modernism. Neither of them is a fashionable figure in our current genealogy of modernism. But their effectiveness cannot be ignored. Indeed, it was Nicolson who in his just-futurist novel *Public Faces* (1932) (the title a citation from Auden) places the outbreak of the Second World War in June 1939 and makes its cause the dropping of an atom bomb, thus again reminding us how fully the destructive potential of nuclear fission was recognized in this period. Nicolson praised and tried to explain Joyce's techniques for a radio audience in December 1931 in a series 'The New Spirit in Literature' which includes his discussions of Woolf and Auden too. The passage he chose to quote and to rewrite bears on exactly the questions of substance, perception, optics and emptiness that the physicists Thomson (Lord Kelvin), Oliver Lodge and Eddington had recently been expounding in radio talks. And he chooses a passage where waves are part of the subject matter. In *The Listener* version of his talk Nicolson explains:

Joyce wants to describe a man walking along the beach. On the inferential system the experience of this man would be recorded more or less as follows:
When Stephen reached the beach the tide was already nearing high water mark. The sea stretched before him in a line of blue and silver melting into a diaphanous distance. The incoming tide deposited on the shingle a fringe of jetsam – an old green kettle there, a bundle of wet straw, and here an old rusty boot. 'How strange,' he said to himself, 'are the tricks one's vision plays! I *feel* colour, but I know about, and therefore *deduce* solids. I feel the blue and silver beauty of the June morning. Yet I recognise that old boot as a boot because I remember having seen boots before. It is the same with sound. One's perceptions are an interchange between cognition and memory.' Thus musing, the man walked along the beach. His shoes crunched on the dry shingle . . .
and so on. Here you have the direct, continuous, conscious narrative of the old-fashioned technique. Now here is how Joyce treats the same experience:
Ineluctable modality of the visible: at least that if no more thought through my eyes. Signatures of all things I am here to read, sea-spawn and sea-wrack, the nearing tide, the rusty boot. Snotgreen, blue-silver, rust: coloured signs. Limits of the diaphane. But he adds: in bodies. Then he was aware of them bodies before of them coloured. How? by knocking his science against them, sure. Go easy. Bald he was and a millionaire, maestro di color che canno. Limits of the diaphane in. Why in? Diaphane, adiaphane. If you can put your five fingers through it it is a gate. If not, a door. Shut your eye and see.

The emphasis on mentalism and on multiple representation without hierarchy is there in Joyce or Woolf as much as in Eddington, or Jeans's *Mysterious Universe*:

We need no longer discuss whether light consists of particles or waves. On
our days of thinking of it as waves, we may if we please imagine an ether to
transmit the waves, but this ether will vary from day to day . . . we need not
discuss whether the wave-system of a group of electrons exists in a three-
dimensional space, or in a many-dimensional space, or not at all. It exists in
a mathematical formula: this, and nothing else, expresses the ultimate reality, and we can picture it as representing waves in three, six or more
dimensions whenever we so please.

For Jeans a permissive scatter of representations still depends on a
final authority – that of mathematics. Eddington went past this
point, chafing Jeans for assuming that mathematics was God, because Jeans was a mathematician.

The writer of the passage on modernism that I quoted first,
Charles Hooker, a popularizer of mathematical theory, refuses to
discriminate between 'efficient diagram' and 'accurate model of reality'. Perhaps there is a hint of a sneer in that evenhandedness, a
suggestion that we are here dealing anyway with whimsical representations. Modernist physics is not bound to single models or diagrams. Indeed, in 1927 Niels Bohr first put forward the theory that
observations of atomic or subatomic systems require more than one
simultaneous concept to be in play, wave-particle duality ('waves
carrying energy may have a corpuscular aspect and . . . particles
may have a wave aspect'): which of the two is more appropriate
will depend on what is to be explained. 'The whole trend of modern
scientific views is to break down the separate categories of "things",
"influences", "forms",' writes Eddington in the Introduction to *The
Nature of the Physical World*. Both 'nature' and 'the physical' are
shifting their shapes.

Elsewhere in *What is the Fourth Dimension?* Hooker uses another
disparaging term for such theorists: they are 'modern highbrows'.
The link is taken for granted between the concept 'modern' and the
expression 'highbrow' (pretentiously intellectual, against common
sense, by implication *unsound*). That's a widespread response in this
century to the idea of modernity, in whatever zone of thought or

experience it appears. It is also a very odd one. Why should the modern be seen as 'highbrow'? It is as if most people construe themselves as living in the past, usurped of the present – which has been occupied by highbrows, ultra-modernists. The resentment is directed against scientists and the implications of their theories alongside other 'difficult' modernist writing. Three years later Norman Edwards, the editor of *Popular Wireless*, joined in an attack on the BBC for its 'highbrow' talks policy – a debate that continued throughout the next eighteen months and which *The Listener* lengthily indexes under 'highbrowism'. Edwards was critical of Eddington on the grounds that his lecture on 'Matter in Interstellar Space' 'must have puzzled hundreds of thousands of listeners' and as his knock-down argument Edwards tells how he 'once heard a lecturer who was broadcasting refer to a four-dimensional continuum'. (*The Listener*, 23 April 1930, has an editorial entitled 'Are talks too highbrow?' followed by Norman Edwards's article.)

Other correspondents in the controversy over 'highbrowism' pointed to the sheer eclecticism of the new audience of radio (or as they would more often have said) wireless listeners. To whom should an expositor speak? That sense of an evanescent and unfocused crowd alternated dauntingly with an insistent single listener receiving speech intimately and at home.

The debate about radio talks has wider implications for literary modernism since it brings out the degree to which the idea of the reader had become unsettled, both socially and formally. As she wrote *Between the Acts* at the end of her life, Virginia Woolf was desolated by the lack of 'an echo' – she sensed that she was writing into a void, no longer to a sharing readership. In such circumstances irony itself may become void, its angle of deflection without poise. Instead, overhearing, or being alongside, hearing cows bellowing, gramophone music, gossip, megaphone, the play: these are the forms that connection may take. Woolf disperses meaning and story through many voices here. With all her distrust of patriotism she yet needs to find a place within community, both the country community within which she is sheltering as the war comes on and the broader community (if such there be) of England and of Europe.

She finds it in unnamed and anonymous voices and their
undersong.

... At Larting no one goes to church ... There's the dogs, there's the
pictures. It's odd that science, so they tell me, is making things (so to speak)
more spiritual ... The very latest notion, so I'm told, is, nothing's solid ...
There, you can get a glimpse of the church through the trees.

Sounds matter, sometimes more even than signification. Thoughts
chime and decay, rhyme momentarily holds unlike together, time
seizes remote past and present instant in a single glance, community
pulsing in atomized presences. Here rhyme, rhythm, wave motion,
pulse, are one movement, while against them is the stultification of
the war about to foreclose possibility. The moment of the book is an
afternoon in mid-June 1939, the dry summer before the onset of
the war. Isa seeks – and cannot find – a cure in books; her list of
hoped-for cures ends with 'Eddington, Darwin, or Jeans'. Within
the work Isa comforts herself with rhyme, and rhyme also spreads
through the thoughts of the audience assembled for the pageant.
Assembled they may be, but *'Dispersed are we, the music wailed;
dispersed are we.'* And the word 'we' scatters through the page a
spoor of rhyme words, none given precedence above the others:
tree, company, me, nursery, see, agony. 'Then the music petered
out on the last word *we.'*
   Rhyme is the cling and turn between individual and communal.
Woolf had heard how the rhythmic underconsciousness of language
could be abused, and founder. In September 1938 she and Leonard
heard Hitler speak at the Nuremberg rally, live on radio. She wrote
in her diary:

Hitler boasted and boomed but shot no solid bolt. mere violent rant, & then
broke off. We listened in to the end. A savage howl like a person excruciated;
then howls from the audience; then a more spaced and measured sentence.
Then another bark. Cheering ruled with a stick. Frightening to think of the
faces. & the voice was frightening. But as it went on we said (only picking a
word or two) anti-climax.

The few words understood speak anticlimax but the undertow of sound speaks volumes of what is to come in the years about to begin. 'Frightening to think of the faces. & the voice was frightening': that expansion of *voice* divorced from sight is the characteristic new experience of the 1930s.

For us now the visual is the public medium but for them it was, newly, speech and sound on their own. At the start of this essay I emphasized the intimacy of wireless, merging so freely with domestic life and with solitary thought. On radio that intimacy is the medium also of public and communal experience. No secure boundaries prevail between private and public. Through the airwaves come complex ideas and propaganda, music and rant: the rhythms of sound energizing the connections between everyday experience and long arcs of meaning, brouhaha and prophecy. No wonder then that in the baleful mid-1930s when Eddington sought an image for the end of the physical world, he imagined it as 'one stupendous broadcast'.

Dome Days:
Buckminster Fuller in the Cold War

ALEX SOOJUNG-KIM PANG

No other object so well symbolizes the American commune of the 1960s as the geodesic dome. From 1966 until the mid-1970s, the dome was the physical embodiment of countercultural possibility, the empty vessel in which a new, egalitarian society could be contained. Not only was the dome the shell from which the counterculture would hatch, but dome building provided craft skills and built a feeling of solidarity among communards. The Red Rocker commune in Colorado epitomized the new linkage between dome and society building. In *Domebook Two* (1971), they explained that they chose a sixty-foot-diameter dome as the centre of their community because 'we wanted our home to have a structural bias against individualism and for communism . . . There's no point in building revolutionary structures to shelter reactionary life-styles.' Equally important was the experience of building the dome: 'We tried very hard to keep an oppressive division of labor from developing and we did pretty well: all the Red Rockers are domebuilders . . . [making] domes and Revolution together.' Several thousand miles to the north, however, another and more extensive set of geodesic domes had for over a decade been put to radically different purposes. In northernmost Canada, stretched out over a three-thousand-mile line, was the Strategic Air Command's DEW line, the nation's first line of early warning against a long-range bomber attack from the Soviet Union. The DEW line, designed by MIT's Lincoln Lab and built between 1954 and 1957 by Western Electric, and managed by IT&T, consisted of hundreds of radar installations, a chain of large and small facilities with housing, offices, tracking stations and radar dishes. Standing alone on the Canadian plain, the most prominent visual elements were the covers that housed the radar and kept them safe from the bitter winds and snow: the radomes, large geodesic domes made of fibreglass and plastic.

In fact, years before hand-made domes of rough plywood, Plexiglas and scrap metal mushroomed across America, the dome had served as a tool and symbol of Cold War corporate and military America. In the 1950s the Air Force and Marine Corps took geodesic domes to the front lines of the Cold War, using them in radar pickets and integrating them into beachhead invasion plans. The Department of Commerce used them as pavilions for American exhibits at international trade fairs. To fair organizers, if not to millions of Afghans, Poles, Japanese and others, geodesic domes became 'tangible symbols of progress' dramatizing 'American ingenuity, vision, and technological dynamism'. In fact, the dome was Cold War technology *par excellence*, and in effect an icon of American industry's power to transform nature, a structure that promised an American patriotism based on consumer values – ironically, a symbol of everything that commune members rejected when they built domes in the late 1960s.

Geodesic domes are complex spherical icosahedra, solids made of thirty equilateral triangles curved into a spherical shape. Each of the thirty triangles can be subdivided into sets of smaller triangles, producing a structure that is increasingly complex and spherical; that complexity is expressed as the variable v, or frequency. Hence, a dome whose icosahedral triangles are divided in thirds (into nine smaller triangles) has a frequency of three (3v). (v also expresses the number of different sizes of struts or triangles needed to build a dome.) Domes are lightweight, have a high surface-to-area ratio, and when assembled properly are extremely strong. The dome's inventor, R. Buckminster ('Bucky') Fuller (1895–1983), was a native of New England, a descendant of Margaret Fuller, a graduate of Milton Academy and Harvard dropout, who served in the Navy during the First World War and began work as an inventor, writer and lecturer in 1927. A basic patent on the geodesic dome was filed in 1951 and granted three years later. It was his greatest intellectual and commercial success. His earlier inventions, most notably the Dymaxion car and house, were never commercially produced, and his later proposals for floating tetrahedral cities or domes to cover Manhattan never left the drawing board. His many books remain peculiarities to all but the most loyal of his admirers, their style signalling that they are the work of a visionary who has invented his own language to convey what he has seen and imagined. Only the dome moved from the drawing boards and circles of devotees into the real world. The failure of most of his inventions has usually been considered proof of Fuller's being 'ahead of his time', a lone inventor who struggled against corporate America. However, this picture is completely wrong. Before he became the iconoclastic prophet of the 1960s, the counterculture's technological guru, the success of the geodesic dome in the 1950s allowed Fuller to be portrayed as the last of the great Yankee engineers, a combination of Henry David Thoreau and Henry Ford. This role could easily accommodate any amount of visionary enthusiasm side by side with hard-headed business skills. Indeed, he worked for Phelps Dodge and *Fortune* magazine in the 1930s, was consultant to companies like Ford in the 1950s, and did some of his best work for the Marine Corps and the Strategic Air Command. Far from being a solitary inventor unconcerned with financial gain or the grubby realities of business, he was well versed in patent law, kept a watchful eye over the use of his inventions and ideas, and was an aggressive manager, founding several consulting companies that provided him with collaborators and connected him with professional societies, industries and universities.

No inventor ever works without collaborators or in isolation from society, and Fuller was no exception. His innovations were produced within a network of consulting firms, industries and schools that provided him with projects, resources and labour. The early years of the Cold War saw Fuller at his creative peak, thanks in part to the support of that network. His headquarters from 1946 was the Fuller Research Foundation in Long Island. By 1951 it had branch offices in Chicago and Montreal (the latter created to give Fuller access to aluminium alloys, which were rationed in America), operated by students whom he met at the Chicago Institute of Design in 1948. Fuller also created two other companies: Geodesics, Inc., which handled government and military contracts, and Synergetics, which was created to promote commercial development. Most of Fuller's more advanced work, however, was conducted in the form of short projects at architecture schools, using materials donated from industry and labour from student volunteers. As copies of Fuller's lecture schedule in the early 1950s and a master list of nearly all his speaking engagements from 1928 on show, Fuller gave only a handful of talks in the first years after the war, but his schedule took off after 1948. In 1949–50 he devoted sixty days to fourteen engagements; during 1951–2 he spent ninety-one days giving twenty-three courses or lectures; and in 1952–3 he managed to give an incredible 295 days to thirty institutions, delivering keynote addresses, serving as a judge or critic at design competitions and holding month-long visiting professorships at architecture schools. (Fuller's pedagogical activities in the late 1940s and 1950s are richly documented, thanks in part to Fuller's own habit of sending large numbers of xeroxes of his personal papers to architecture school libraries around the United States.)

Speeches and seminars gave Fuller publicity, but the visiting professorships were infinitely more important.* Fuller's companies were too small and under funded to conduct serious research and build prototypes; temporary professorships gave him an operating base, students' labour, and high-tech materials that were essential for turning ideas into inventions and drawings into prototypes. Fuller visited MIT and North Carolina State every year between 1949 and 1956, and paid several visits to Minnesota, Princeton and Cornell. These regular visits allowed him to develop a community of seminar students and pursue projects at a single institution over several years. The schedule of each visit was carefully planned. About two dozen

---

* This section draws on the Fuller Miscellaneous Papers, Rotch Library, Massachusetts Institute of Technology; President's Papers, AC 4, MIT Archives; School of Architecture Papers, 77–53, MIT Archives; Henry Kamphoefner Papers, North Carolina State University; Fuller miscellaneous papers, Environmental Design Library, University of California (Berkeley); Robert Marks Papers, College of Charleston and the 'Fuller-World Fellow in Residence' files, University of Pennsylvania Archives.

students would study with him. They began with a week of marathon lectures on Fuller's philosophy and designs. The second week would be spent 'tooling up' to build a prototype dome, trying out new construction materials or methods, designing the struts, hubs and panels that would form the dome, and drawing or cutting patterns. In the last two weeks, the parts would be produced, the dome would be assembled, and summaries and assessments of the project would be written to guide future work.

Fuller's visits were expensive. Not only did the colleges pay his salary, they had to cover bills for building materials, many placed as rush orders during moments of inspiration. Further, as one college dean warned a friend, he could be a bad influence: 'Many of the students will fully reject all other teaching they have had in your school because of their experiences with him, and I think you had better expect that.' More deeply, Fuller's unorthodox inventive method created problems of ownership and credit that he was anxious to resolve. Before classes began he made his students promise to 'protect my proprietary rights', and required schools to 'waive any "shop rights" claims upon my patents or design developments'. This was essential, he argued, because 'I cannot afford to bring my lifetime work to them and expose my technical frontiers to loss of control and loss of initiative.' He also claimed ownership of any dome students designed and built under his supervision, and downplayed the importance of their contributions. 'It must be remembered,' he argued in the course of a dispute with dissatisfied students at Washington University, 'that the dome was manufactured and erected . . . only because I had an experience-fertilized teleological design back-log' that guided their work. The students, in other words, had only built what he had already imagined. Even more remarkably, he claimed ownership of any work by students inspired by him but conducted independently, declaring that a student-authored publication describing his geometric ideas 'represents *my own work* and effort of a third-of-a-century's continuous processing'. These incidents show that Fuller believed that any work done on domes or based on his ideas about industrialization, geometry or any other subject, was *his*; since no student was capable of going

beyond him, any work that any student did was *a priori* either
owned by Fuller or stolen from him.

Still, many schools felt that Fuller was worth the risk, at least in
small doses. Why was that? For two decades he had been a marginal
figure in the architecture and design world. What accounts for his
sudden popularity in the late 1940s? The answer lies in the challenges the profession faced in the early Cold War. For a few years
after the Second World War it enjoyed an explosion in private and
public construction spurred by the release of wartime controls on
building materials and credit. Cruelly, by 1950 it was all threatened
by the twin threats of nuclear war and permanent war mobilization.
Atomic destruction was clearly the worse of the two, but a controlled economy geared for wartime production promised a less total
but rather more tangible disaster. The problem was to find a way to
respond to national security challenges without sacrificing civilian
production.

The solution, according to many architects, was 'dispersal'.* The
threat of nuclear destruction had rendered the traditional city inde-
fensible and obsolete, its advocates argued; the only way to protect
American people and industries from massive destruction was to
break up urban and industrial concentrations and rebuild America
as a set of small towns, hardened against nuclear attack and separated by green belts and fire walls. Like the origin of the Internet in
the defence-led ARPAnet project for a communication system too
centreless to be disrupted by enemy attack, dispersal showed that
the apparently radical goal of decentralization could well be a military imperative. Far from being the interest of a fringe in the profession, these proposals were discussed by American Institute of Architects presidents, leading practitioners and academics in places like

* Major archival sources on dispersal include the National Security Resources
Board, Record Group 304, National Archives (Washington, DC); *U.S. National
Security Resources Board Press Releases*, Volume 1, 1948–1951 (unpublished
MS, UC Berkeley Government Documents collection, n.d); Clarence S. Stein
papers, Collection #3600, Cornell University Archives; Edmund Purves Papers,
Library of Congress; Norbert Weiner Papers, MC 22, Massachusetts Institute of
Technology archives; Bemis Foundation Papers, MC 66, MIT Archives.

the *Architectural Forum* and AIA annual meetings. The idea had been discussed by strategic thinkers and scientists immediately after the Second World War, but they treated it only as a tactical option in the national defence game. Architects and urban planners, however, talked about really *doing* it, implementing a vast construction programme with immense political and economic consequences. A programme that had as its aim the breakup of cities, industrial concentrations and transportation nexuses was nothing short of a revolution; and carried out by the federal government as a crash programme, it would have required the creation of a command economy, suspension of basic civil rights and a regimentation of society that would have rivalled Stalinist Russia. Thus dispersal's advocates in the federal government, the architectural profession and planning communities worked hard to show how dispersal could be done without government expansion or intervention, economic dislocation or sacrifices in the national standard of living.

Supporters claimed that dispersal could be done economically, would have political benefits and would improve America's quality of life. They began by arguing that suburban expansion and factory relocation from cities to small towns had already got it started. If both were simply done more rationally, and new construction in congested cities was relocated into new areas, dispersal could be achieved in a matter of a couple decades. In fact, advocates *insisted* that it be done gradually. In November 1948 the federal government's National Security Resources Board declared itself in favour of 'progressive dispersion . . . built around normal expansion and obsolescence factors, [or] relocation of urban defense plants', and argued that rapid relocation of entire industries was alarmist, rash and uneconomical. If done carefully, dispersal would also be economical and socially beneficial. Tennessee Valley Authority planner Tracy Augur argued in the May 1948 *Bulletin of the Atomic Scientists* that dispersal 'costs us nothing', since 'the building and rebuilding of cities will go on' regardless of the political situation. By eliminating slums and preserving open spaces, Tracy Augur argued, dispersed towns would 'secure a much finer environment for home and work than the average citizen now dares dream of'. They would also eliminate fifth-column activity and agitation, since 'subversive ideas and actions' could flourish only in 'the decay of the city structure in which increasing millions of people spend their lives'. Likewise, Albert Mayer wrote in 1951 that dispersal would turn 'the emergency mood' of the Korean War into 'positive action' by investing national energies rather than wasting them on air-raid drills. A year later, journalist Sylvian G. Kindall argued in *Total Atomic Defense* that dispersal would 'combat crime and corruption by destroying the roots of their evil.' In fact architects placed as much emphasis on using dispersal to improve Americans' quality of life as on improving national defence. Strategists and scientists had considered the economic stimulation, reduced unemployment and improved quality of life created by dispersal to be mere 'incidental advantages', but for architects those effects were of critical importance. Thus city planners Donald and Astrid Monson wrote in *American City* in 1951 that even if dispersal proceeded but war never came, 'the present fearful and awful threat' would be 'turned into so great a blessing that men ... will say that the greatest benefit which flowed from the explosion at Los Alamos was the enforced rebuilding of our urban centers.'

Dispersal would not only bring prosperity: it would also provide renewed access to the mythical American small town, home of the simpler, more honest life and elevating, nation-strengthening eternal virtues. In a time of permanent crisis, Americans were going to declare their faith in the nation by returning to the places that produced and sustained the nation: the last line of defence would also provide strength for the fight. For proof that this was an important subtext, consider the fate of proposals for underground relocation. Between 1945 and 1948, underground relocation was discussed alongside dispersal as an option for national defence, but technical studies by the NSRB suggesting that it would be uneconomical, and the belief that going underground constituted 'hiding out' from Soviet bombers, killed it off. The dispersed small town had great appeal, and could unify opposing groups: industry saw it as a new land of uncontentious and unorganized labour, while the Congress of Industrial Organizations saw it as an escape from 'the

misery which comes from overcrowding in big cities'. Social improvement would follow industrial relocation automatically. It is an irony
worth noting that the dispersal policy influenced the flight of employment from American cities, which in turn helped cause the social
problems that radicalized Fuller's next generation of admirers in the
1960s.

Dispersal offered freedom from the threat of nuclear war, improved standards of living, and the elimination of slums that bred
moral depravity and fellow-travelling activity. What was the catch?
It had to be done privately, with minimal government involvement.
Though it had the authority, the NSRB refused to mandate industrial dispersal, working instead through appeals to corporate heads,
industrial real-estate agents, and factory planners, and indirect in-
centives like tax breaks for dispersed factories. If it were not done
privately, dispersal would destroy democracy even as it saved
America. The paradox of this argument is captured in Kindall's
Total Atomic Defense. In it, he called for zoning boards to outlaw
buildings taller than two storeys, the creation of a super state out of
Massachusetts, Connecticut, Rhode Island, New York and New
Jersey (which he called the 'Area of Utter Destruction'), and relocation of the federal government from Washington, DC, to the Rocky
Mountains. But Kindall opposed plans that forced homeowners out
of their houses. Too little planning and America would remain vulnerable to attack; too much government control and 'we shall destroy democracy by trying to accomplish atomic security. In that
case, to choose between the two forms of destruction, it would be
just as well to sit tight and wait for the A-bomb to come and do its
worst.' Architects feared that government intervention would bring
a return of controls on construction materials and credit, and the
end of the postwar boom in construction; they could hardly promote
dispersal if it led to a Soviet-style command economy. A few weeks
after the Korean War began, *Architectural Record* editors urged readers
to make themselves available to defence agencies, but argued that
'national purposes must be accomplished with minimum restrictions
of our freedoms, and minimum disruption of the civilian economy'.

Architects and builders hoped to find a way to respond to military challenges without sacrificing civilian production, and urged government to leave the construction of new plants and factory towns in private hands. But what private body had the technical skills, managerial expertise and position in the construction industry to oversee dispersal? The answer is not hard to guess. Architects and planners had the experience necessary to plan dispersed towns. They had professional organizations that could coordinate national activities, provide expert advice and solutions to difficult problems, set standards and building codes, monitor the production of building materials and manage relations with labour, builders, finance and governments. Architects used dispersal to define a new service role for themselves in the national defence as providers of the plans that would make America nuclear-bomb-proof, and volunteered to serve as a surrogate for the government in overseeing this vast, voluntary enterprise. Keeping dispersal in the hands of private organizations and the free market was essential: having the government take it over would have represented a massive intrusion of the state in the lives of individuals. Nuclear defence, which was about defence of the American way of life, was too important to be left to the federal government.

Dispersal was a solution to all the key problems facing architects in the Cold War: it was simultaneously a contribution to the national defence, a bulwark against socialism and government overextension and a way to guarantee national security through privately sponsored economic growth. It was also infinitely adaptable to local professional circumstances and ambitions. North Carolina State's School of Design, for example, cooked up a programme that mixed dispersal, regional planning and industrial policy. Industries in the North and Great Lakes would be relocated in rural areas in the Midwest and South, modernized, automated and hardened against Communist nuclear attack, while workers and their families would follow the factories to small cities and towns. The idea generated a great deal of enthusiasm among NC State architects and state officials for a couple years, because it dovetailed nicely with the state officials' desires for economic development and the university leadership's desire for the school to play a role in state improvement.

Fuller's popularity, I argue, came from the fact that his work could be seen as cutting-edge dispersal research. All his work at NC State, for example, was dispersal-related. More generally, Fuller was popular because his work was concerned with the challenges the threat of nuclear war posed to American security and the American way of life. His work and thought, while always unique, resonated with that of the dispersal theorists, and like them he saw in the nuclear threat opportunities for preserving American civilization, raising standards of living and assuring for architects a prominent role in the national security state. Further, he saw clear links between the provision of consumer goods and the military and political triumph of the United States.

What may have fascinated listeners most was the idea that they would not have to choose between guns and butter; butter might even prove more powerful than guns in the Cold War. Fuller was particularly compelling because he could see in the problematic present a future of limitless expansion and opportunity. This is exemplified in a 1949 lecture he delivered at the University of Michigan on the future of small buildings, particularly houses. He identified five forces that would pressure the house into becoming an efficient, mass-produced machine for living. Three were related to the Cold War: permanent industrial and technological mobilization, 'enforced occupation of hitherto hostile equatorial and polar environments' and 'military conquest of the air'. Dwelling units would have to operate everywhere from Siberia to Indonesia, but thanks to permanent mobilization, scientific advances would be applied to housing more quickly than in peacetime, and improved dwellings could be airlifted around the world. The connection between his work and Cold War culture is spelled out in the fourth and fifth forces. The fourth force was the reorganization of 'the industrial complex' around the recognition that 'the key to economic and subsequent political expansion is the *consumer.*' The more consumers there were, and 'the more numerous the consumers' needs and the more frequent their satisfaction,' Fuller continued, 'the greater the expansion.' The combined force of economic mobilization and consumer satisfaction would ultimately 'accelerate total world occupation by total world man in total dynamic enjoyment'. The fifth force, a rapidly rising world population, would provide plenty of consumers for this new industrial complex. In other words, Fuller envisioned an 'efficient' consumer society, feeding on itself and a growing world population, ultimately conquering and transforming the world. 'Total dynamic enjoyment,' he concluded, would ultimately lead to 'total world occupation' by 'total man'.

Buckminster Fuller was thus a very unusual Cold War techno-crat, but a technocrat none the less. What of the dome itself and its history as a technical and cultural artefact? From the mid-1950s geodesic domes were used not as atomic-bomb-proof shelters of dis-persed, automated factories and cities providing 'total enjoyment' to 'total world man', but as shelters for military personnel and equip-ment, radar stations and trade fair exhibits. But even if they di-verged from lines laid down by Fuller's earlier interests, they still occupied the same cultural ground. In Fuller's mind the adoption of the dome by the military and the Commerce Department opened the door to widespread civilian applications, and held the key to America's ultimate triumph in both the 'hot' and Cold wars.

Between 1953 and 1956 Fuller worked with the Marine Corps and Air Force to develop the dome's military applications. To Fuller, this work demonstrated the applicability of the dome to complex technological problems, provided valuable publicity and pointed the way to civilian and commercial exploitation. The geodesic dome's first direct contact with the military appears to have been in early 1949, when a prototype was erected in the Pentagon Garden at the request of the Air Corps.* That same year, Fuller and architecture students at MIT designed an 'autonomous dome' for housing Air Force fighters and crews. Two years later, as noted in his 'Schedule of Lectures and Projects', Fuller and his MIT students designed an eighty-foot 'Arctic hangar for carrier-based U.S. Marine Corps planes'. It is unclear whether these were theoretical exercises or whether Fuller was in contact with officers interested in using the dome. By 1953, however, Fuller was talking to Marine Corps officers

* Information on military uses of the dome comes from material in the United States Marine Corps Historical Center, Washington, DC.

about the dome's potential uses, and shaping student projects accordingly. The Marine Corps' most enthusiastic advocate, Aviation Logistics and Materials Branch director Col. Henry Lane, apparently became interested in geodesics as housing and hangars just after the Korean War. At the time Marine Aviation bases were built in three phases, beginning with tents and ending with permanent buildings. Lane wanted to replace this with a one-phase system in which permanent housing, maintenance facilities and hangars for an entire wing would be prefabricated in America and dropped into combat zones at the first sign of Communist mischief. Such bases would be fully operational almost immediately and realize great savings in time, labour and material. The structures containing these facilities had to be light and sturdy, built from materials that did not attract radar, and mass-produced by industry, preferably without requiring factories to convert from civilian to military production. The dome seemed perfect for the job. It was inexpensive, durable, extremely easy to transport, could be erected quickly with unskilled labour, and its space could be used in a variety of ways. It needed to be manufactured to high tolerances, but the precision labour invested in the factory would be saved in the field.

Fuller broke the Marine Corps research down into four projects. The first, conducted at North Carolina State in January 1954 and co-directed by Fuller and Professor James Fitzgibbon, produced a thirty-six-foot helicopter hangar dome made out of materials donated by Owens-Corning and Union Carbon and Carbide. Fuller next went to New Orleans to direct a group of Tulane students designing and building an eighteen-foot-diameter barracks dome out of cardboard. Both domes were given to the Marine Corps. (Two larger versions of these cardboard domes, designed to Marine Corps specifications, were exhibited at the 1954 Decima Triennale in Milan, where they won gold medals for best pavilion design.) Fuller's second patent relating to the geodesic dome, covering methods for making domes out of cardboard, was based on this work. Once that project was finished, Fuller spent a week lecturing in the South, then travelled to MIT, where he directed a third month-long project, building an aircraft maintenance hangar dome out of plastic.

Marines moving a dome into position. 'New "situation" for the Marines: Fuller's dome may shelter them.' (*Architectural Record*, June 1954, p.24)

Finally, a fifty-foot dome built from magnesium alloy struts donated by Dow Chemical was built at the University of Michigan. These domes appeared to be quite promising. They could be made from easily available materials, shipped flat and assembled in a matter of hours, and could withstand high winds and other stresses. During mock beachhead invasions and other demonstrations, they were easily carried by helicopters and quickly readied for use. They were less vulnerable to bomb blasts and required less heating fuel, promising shorter logistical lines. Equally important, an aircraft wing could be sheltered for almost 90 per cent less than competing structures, and with 97 per cent savings in both weight and man-hours required for construction. The Corps eventually had 300 built, and used them all over the world; however, they never became the replacement for all tents and temporary structures that Fuller and Lane envisioned.

Still, Fuller considered the project a terrific success, as two extraordinary letters to Marine Corps Major George King in late 1955 and 1956 reveal. It was a success, he argued, because it verified two of his key ideas about the place of technology in modern world: the 'Airocean World Plan', developed in 1927, on the consequences of global air travel; and his concept of 'controlled environments' – prepackaged and self-contained cultural and technological units – and their central role in industrial civilization. Fuller's Airocean writings predicted the evolution of an integrated, industrialized global economy linked together by an air transportation system serviced by airports and communities contained in preassembled, self-contained towers airlifted all over the world. These communities represented the 'controlled environments' that Fuller saw as essential foundations for high-tech industrial activity. Scientific and medical research, precision manufacturing, high-yield agriculture and maintenance and service of complex technologies could be conducted only within a technological infrastructure consisting of utilities, transportation and communication facilities, and work and living spaces whose environments could be adjusted and maintained to meet specific needs.

He argued that the Marine Corps' adoption of the dome verified

Helicopter lifting dome. From R. B. Fuller and Robert Marks, *Dymaxion World of Buckminster Fuller* (Anchor Press, 1960, p.203)

his 1927 predictions and laid the foundations for civilian exploitation of his ideas. Industrialized war made logistics an 'integral component of the weapons package'. Marine aviators, gearing up to
fight 'brush fire' wars (Fuller claimed to have coined the term in
1927) in the Third World, required forward technical support services that could only be provided in controlled environments. His
domes were the structures that would provide the 'high standard,
airliftable environmental controls' needed by the Corps, and he described a mock landing staged by the Marines at the 1956 National
Air Show in Philadelphia as the 'first full-scale public witnessing of
the practicality of . . . Airocean World deliverable buildings'. Further, the tools that were developed to win industrialized war could
be applied to winning industrialized peace, for the key to winning
both potential 'hot wars' and the Cold War, Fuller concluded, lay in
effectively transferring technological systems contained in controlled
environments into untamed regions of the world. As he put it in a
memorable phrase in his July 1956 letter, 'the side which has the
superior . . . environmental controls will win.' Developing Third
World countries needed not just simple transfers of technologies
but transfers of entire controlled technological environments. Factories, power systems, schools and hospitals could be produced in the
West and flown into impoverished and overpopulated areas, turning
them into self-supporting, industrialized nations overnight. The Marines, in other words, had developed a 'double-barreled Geodesic
weapon' that could win both hot and cold wars. Through their
efforts, he told King in September, the United States could not only
turn back Communist aggression, but turn 'the world's scientific
and industrial potentials to the greatest common advantage of man'.
Ultimately, Fuller predicted, the Marine Corps' adoption of the dome
would pave the way to a global adoption of 'the American economy
and the democratic processes which provide the synergetic strength
of the USA'.

In fact as Fuller wrote these words the dome was beginning a
second career as a propaganda tool, helping spread the gospel of
American democracy and capitalism behind the Iron and Bamboo
Curtains. Fuller was not as deeply involved in this part of the dome's

development and use, but its history shows how the dome could be made into a political symbol for a culture in which the categories of military and civilian, economics and politics, and political freedom and consumer plenty were dissolving. Exhibits 'selling the American way of life' to people in the Third World and behind the Iron Curtain recombined popular culture and national security, presenting America as a capitalist paradise of cars, appliances and frozen foods, and offering consumer goods as the reward for embracing American-style democracy. The geodesic dome itself became part of the exhibit as the quintessential American invention, and with it Fuller was legitimized as the quintessential American inventor, an iconoclast who combined vision, practical mechanical ingenuity and an unswerving independence of spirit.

International trade fairs and expositions were battlegrounds in the Cold War, marketplaces of ideas and ideologies. Communist nations, especially the Soviet Union, East Germany and Czechoslovakia, began using trade fairs as propaganda platforms in the late 1940s, and in response, between 1954 and 1956 the United States constructed the instruments for creating and carrying forward the American message on to the exhibition battleground. The White House organized exhibits until the Commerce Department opened its Office of International Trade Fairs (OITF) in 1955 to oversee US exhibit planning. The following year, Congress, declaring that musical groups, sports teams and trade fair exhibits had become 'permanent weapon[s]' in the Cold War, passed legislation giving the President, USIA and Departments of State and Commerce authority to coordinate American activity in a variety of international events.* The OITF handled planning, organization and delivery of pavilions and exhibits. A Trade Fair Subcommittee, operated out of the National Security Council and consisting of the representatives of the Departments of Commerce and State and of the

* Information comes from papers of the Office of International Trade Fairs, Record Group 489, National Archives (Washington, DC), and correspondence in the collections at MIT, NC State and the University of California (Berkeley). Other material comes from US Department of Commerce and US Information Agency annual reports, 1954–60.

USIA (and from 1959, the CIA), designed the principal exhibits and set the rules for companies, organizations and trade groups who wanted to join the delegation or contribute displays.

The trade fair exhibits created under this system were a remarkable blend of unswerving patriotism, innocent pride and unselfconscious kitsch. The first American exhibit was unveiled at Bangkok in December 1954. United by the theme 'the Fruits of Freedom', 140 firms exhibited everything from farm equipment to automobiles to 'See Yourself on TV' displays. Over the next several years, organizers grew especially fond of three exhibits. The first, 'America at Home', featured a furnished house inhabited by 'actors impersonating an American family – husband, wife, and two children', who 'demonstrated what their standard of living meant: using the kitchen appliances, tuning the television set, and so on'. The second, Main Street, USA (copied from Disneyland), featured 'stores, a church and . . . a kindergarten with teacher and children'. Finally, the 'typical American supermarket' filled a walk-through exhibition hall with canned, boxed and frozen foods, products that exhibition organizers thought would be especially appealing to visitors in the Third World.

The geodesic dome was an integral part of this propaganda front in the most aggressively contested areas of the Cold War world. The first dome appeared at the International Trade Fair in Kabul in 1956. It was fitting that Fuller's technocratic vision should find its way on to the trade-fair grounds of Afghanistan, a country that had long been an object of imperial machinations and cross-cultural contact. In May, a mere five months before opening day, OITF officials received reports of planned Soviet and Eastern Bloc exhibits: the Soviets were going all out, intending to show the Afghan public the benefits socialism had brought to their northern neighbour. The Chinese were likewise planning a large exhibit. If they didn't act quickly, an American pavilion would be shunted off into a corner of the grounds, or shut out entirely. It was too late to design a permanent building, and the portable aluminium and canvas pavilion that was used in other fairs was in need of repairs. OITF director Harrison McClung first planned to order another portable tent, but OITF architect Jack Masey complained that tents leave 'much to be desired in terms of aesthetic appeal', and it would be inadequate to the task of countering 'what seems like very severe Iron Curtain participation'. A tent would suffer in comparison with Soviet and Chinese buildings and therefore suggest inferiority of the American system to socialism. Instead, Masey suggested, they should send a prefabricated building. Several companies were making them, and Buckminster Fuller might be able to design a dome in time. Why they chose Fuller is not recorded, but by the end of May Synergetics had a contract to design and construct the pavilion for the Kabul fair.

After only a week, the North Carolina Geodesics office completed plans for a dome with a 100-foot diameter, and was overseeing its production. Made of 480 aluminum tubes, 2,000 bolts, 143 hubs, and 1,300 pounds of nylon skin, it was the largest geodesic dome yet built. The dome was designed and built so quickly that OITF engineers were suspicious that it might not stand, and ordered a review of the structure. It was shipped out of Raleigh in mid-July and was erected on site by four local workmen in two days. At first reaction was mixed. Robert Warner, who had been suspicious of the dome all along, and some of the other businessmen who contributed to the exhibit, were underwhelmed. Warner complained to McClung that 'our pavilion was much too small', and because it 'was translucent, its small size was emphasized, particularly during the evening hours'. An executive with Singer thought the dome 'was a very poor show', but admitted that it was popular with the visitors. In fact, it was *overwhelmingly* popular with visitors, and that probably turned the tide in official opinion. According to *The Nation* in March 1957 and the *Architectural Forum* in November 1956, Afghan visitors climbed up the dome and had to be removed by the police, the King asked to be given the dome as a gift and even 'Russian engineers kept busy with sketch pads at the U.S. exhibit'. The American embassy proudly reported that 39 per cent more people visited the dome than the Chinese exhibit, and 35 per cent more than the Russian. The Communists' large and ponderous exhibits were no match for the smaller, more pleasing dome.

Virtually overnight the dome blossomed into a symbol of American technological ingenuity and hence the vibrancy of American capitalism. In an OITF letter sent to business contributors, Tom Hall Miller reported that 'the outstanding appeal' of the exhibit was 'due to the combination of the dramatic Geodesic dome . . . the cinemascope theater' and the use of local interpreters. He told Fuller, 'In my opinion the geodesic dome is the most dramatic structure we have utilized in this part of the world.' According to an OITF press release, 'The American building itself turned out to be U.S. Exhibit #1. Covered with a translucent plastic-coated nylon which glowed at night, the geodesic dome called dramatic attention to American technological progress.' Transcripts of the speeches made by pavilion guides show that visitors were treated to the same interpretation. 'The very dome which houses these exhibits', they were told, was a

demonstration of the degree of industrial progress attained in the US . . . It emphasizes the marriage of aesthetics and technology – two very vital phenomena which are symbolic of a people who believe that only through peace can there be progress.

The dome was also a big hit among American officials because it was easy to erect, covered a large floor area and was cheaper than permanent facilities. Within two years, according to *The Nation* (March 1957) and the *New York Times* (January 1958), eight of the 'catenary geodesic tents' (of 100- and 114-foot diameters) were being used by the Commerce Department to challenge the 'Soviet "economic offensive" in India and other Asian and African countries', housing exhibits in Bangkok, Tokyo, Rangoon, Tunis and Casablanca. A ninth was built to house the US Information Agency's 'Circarama', a theatre that surrounded the audience with scenes of the Grand Canyon, New York Harbour and Americans at work in shopping centres, ranches and ports. (Circarama was originally developed for Disneyland in 1956; the USIA's version was visited by six million people in ten cities between 1958 and 1962.) By 1960, government-organized pavilions, featuring a mixture of OITF, corporate and USIA exhibits, had repeated this message at ninety-six fairs, and *Business Week* (as early as January 1956) claimed that upwards of fifty million people 'from Addis Ababa to Jakarta' had 'crowded into U.S. pavilions to wonder at atomic power exhibits, voting machines, [and] electric trains'. The domes were used most often in the Third World, both for ideological reasons and because fairs and exhibitions in Western Europe tended to be more specialized, open only to businessmen, and have permanent exhibition facilities. Fuller applied for a patent on the tent domes in March 1957. It was granted two years later, his third patent derived from government work.

The dome reached its zenith in two trade fairs behind the Iron Curtain. The first was the 1957 Poznan trade fair in Poland, covered by *Life* magazine in a July photo essay. The Commerce Department sent 'a geodesic dome packed with things the average American can buy to eat, wear or use to make work easier and leisure more pleasant'. The *Life* article was filled with pictures of anxious and awed Poles deserting the Soviet exhibit, which focused on heavy machinery, and flocking to the 'U.S. wonderland . . . filled with hi-fi sets, dishwashers, automatic ironers, air conditioners, jukeboxes, frogman flippers and power tools for the home handyman'. They jammed food counters, mobbed Coke machines and nearly rioted when frozen food was given away after a freezer broke down. Children flocked to a 'toyland' filled with stuffed animals, envious families filed past a $17,000 'U.S. model home' and Polish women watched demonstrations of ready-mix cake 'as if it were a magician's performance' (or so the *New York Times* crowed). The triumphant message, *Life* concluded, was that 'the United States system has an overwhelming concern for the individual'.

Two years later a second dome was sent to Moscow. There, corporate and government interests and actions were seamlessly joined, both under and in the dome, to produce an exhibition that stretched the conspicuous consumption as propaganda strategy of previous exhibits to its limits. Originally planned as a private venture, the $3 million fair was taken over by the government and corporate subscribers when its original organizers dropped out. Architect George Nelson, a longtime friend of Fuller's, was made exhibition director

and quickly filled the dome, a second hall, and an outdoor display
area with a variety of dazzling exhibits. In addition to the standard
agricultural, medical, scientific and technical exhibits – IBM contributing a computer, RCA a television studio and NASA a model of
the Explorer VI satellite – a number of companies volunteered displays introducing American life to Russians. Whirlpool and other
appliance manufacturers shipped over model kitchens (the Soviets
complained that they had shipped too many over), and one was the
site of the famous 'Kitchen Debate' between Nikita Khruschev and
Richard Nixon. Eight fashion experts were dispatched to give advice
on cosmetics to Russian women, who could buy them at a full-scale
drugstore and supermarket. Finally, in a twist on traditional
American practices in which Indians and other colonized peoples
were put on exhibit for curious whites, the R. T. Davis family of
Milburn, New Jersey, was chosen as the Typical American Suburban
Family and put on display in a ranch house furnished by Macy's.
Charles Eames designed and produced a film showing 'things which
were too big to bring to Moscow, such as dams, highways and
cities, housing, schools, etc.' It was actually several films, projected
simultaneously on seven screens 'to establish credibility'. Entertainment was provided by Ed Sullivan.

Unlike previous domes, which were made to order for the Commerce Department, the 200-foot-diameter, gold-anodized, aluminium
dome used in Moscow was part of a commercially available line
recently launched by the Kaiser Aluminum Company. Henry J.
Kaiser was a fitting partner to Buckminster Fuller. A tireless entrepreneur, Kaiser had erected a 144-foot dome less than six months after
the Kabul fair, made of aluminium panels and bolted together, at
his Hawaiian Village resort at Waikiki and, so *Business Week* reported in February 1957, was planning to mass-produce
auditorium-sized domes for '[every] city of over 10,000'. In May of
the following year a *Business Week* cover article on Fuller cited
Kaiser's entrance into the dome market as proof that the dome had
become commercially as well as politically attractive. Kaiser was
not the first company to use a dome – The Ford Motor Company
had built a 93-foot dome in Dearborn in 1952, and Union Tank Car

Company was building a 384-foot dome – but it was one of the first large firms to produce domes for the market, rather than for its own consumption. Other businesses swelled the ranks of firms taking out licences on the basic and applied dome patents: their numbers rose from eight in 1955 to thirty-one in 1956, seventy by 1958 and a hundred by 1960.

But the dome's greatest success at the end of the decade was not as a *product* of capitalism but as a *symbol* of capitalism. Cultural perceptions were just as important as the dome's technical success in the trade fair wars. To American planners and commentators, the values it drew on as a design solution seemed to embody the universal appeal of the American way of life. Supremely portable, light and durable, it went about being impressive by being, let's say, really neat – desirable as soon as you knew that such a thing existed. According to Robert Marks, domes filled with the latest consumer goods and technologies were 'tangible symbols of progress', and 'better propaganda than double-meaning speeches'. David Cort, an admirer of Fuller's and critic of the 'refined bestiality' of US exhibits, contended that trade-fair battles were being 'won by the dome, not by the merchandise inside'. In his March 1957 *Nation* article he argued that the dome was 'as typically and individually American as the Model T Ford car, the simplest and best way to be a difficult thing' and an embodiment of 'the continuous American revolution of fresh, untrammeled thinking'. Fuller biographer Alden Hatch echoed the sentiment in 1974, calling it 'the only uniquely American contribution to architecture since the skyscraper'.

Soon Fuller too came to be described as 'typically and individually American'. Writers noted his relations to the Transcendentalist writer Margaret Fuller, his roots in 'old New England' and his short career in the Navy. *Business Week* praised his 'uncompromising iconoclasm ... independence, and stubborn integrity', and compared him to Charles 'Boss' Kettering and Frank Lloyd Wright. By 1964 a *Time* magazine cover story on him described him as 'a throwback to the classic American individualist, a mold which produced Thomas Edison and [New England writer Henry David] Thoreau'. In fact, everything about him – his visionary genius, his ambition, his hectic schedule, his long and happy marriage – was made a symbol of America, forever building, improving, imagining, on the go, yet never forgetting or betraying its basic principles and beliefs. By the time it was clear that the dome-building bonanza predicted by writers was not going to materialize, Fuller had moved on to other things; once the dome was a cultural success its commercial success became unimportant.

But Fuller's life and the geodesic dome had been intimately linked for over a decade. The association between him and this particular architectural shape had been permanently forged. (Researchers in the 1980s working on the structure of Carbon-60, realizing that the carbon atoms were arranged in the familiar geodesic form, inevitably dubbed their discovery a 'Bucky ball'.) And, as a defining achievement, an engineered shape proved to be more open to competing claims and understandings than his 1950s audience might have expected. Open to claims from the legal point of view to begin with, Fuller had to be unusually expansive in defining his own rights and restrictive in acknowledging others' precisely because each dome had many contributors, each of whom could stake a claim to ownership. Likewise, the geodesic dome's cultural meaning was not fixed by Fuller alone, nor by any single person in any single place through any single act. Its politics were created through commentary about the dome by journalists, critics and students, the use of the dome by the Marine Corps, Strategic Air Command and Office of International Trade Fairs, and finally by Fuller's own writings and his conscious efforts to present the dome as the solution to the technical and political problems of his patrons. The technology and the society in which it existed were for Fuller co-productions, each drawing on and supporting the other. The geodesic dome depended for its success on the proper ordering of allies, agreement among a circle of men and negotiation over what constituted its proper use.

Thus even during the dome's heyday in the late 1950s, when American writers turned it into an emblem of American capitalism and plenty, some of the visitors who filled domes and filed past the exhibits of frozen foods and automobiles were making very different sense of the structure. Fuller reported that 'the Royal Afghanistan

Ambassador undoubtedly saw in the Geodesic structure' erected in Kabul in 1956 'a realization . . . of the age-old mobile efficiencies of their unique and primitive' technology. (It should be noted, however, that the Ambassador had an engineering degree from Cornell.) David Cort reported that visitors in Kabul 'came inside, fell on their knees and prayed', while others 'said that it was merely ancient Afghan architecture . . . [using] the same universal principles as the nomad's yurt, made of interlaced saplings and sheepskins'. Visitors in Bangkok and Osaka were reported as claiming that the dome was 'our own traditional architecture, industrialized'. Those visitors showed that different groups could reinvent the dome's symbolic meaning, detaching the cultural linkages that one group gave it and substituting their own. In this, they anticipated the history of the dome, and of Buckminster Fuller, in the 1960s, when the commune movement, with their building manuals *Domebook* and *Domebook Two*, would detach the dome from its Cold War foundations and carry it, along with its inventor, away.

Making the Match:
Human Traces, Forensic Experts and
the Public Imagination

ANNE JOSEPH
AND ALISON WINTER

There had never been an investigation like it, and the future of genetic manhunts was now being studied by lawmen from all over the world. The revolutionary murder hunt for the footpath killer had resulted in the blooding of 4,583 young men, the last being Colin Pitchfork of Littlethorpe, whose DNA did indeed provide a perfect match to the genetic signature left by the slayer of Lynda Mann and Dawn Ashworth.

The Blooding, 1989

'There had never been an investigation like it.' Novelist Joseph Wambaugh, a former Los Angeles police officer, came to this conclusion about the impact of the new technique of genetic profiling, or 'fingerprinting', on the tracking of human beings. His tale of its application in the investigation of two rapes and murders in 1980s Narborough is one of the many celebrations of the power of genetic fingerprinting and of the inexorable progress of science in identifying human traces. At the end of this century, genetic fingerprinting is emerging as the ultimate human identifier. From the smallest, the seemingly most insignificant of traces – a few hairs, a sample of urine, semen or saliva – genetic fingerprinting promises to identify a single human to the exclusion of everyone else. Its allure lies in just this characteristic: minuscule residues of peripheral human productions are supposed to contain the essence of an individual's identity. But from where did the conventions come which give genetic fingerprinting its powerful resonance? The prestigious place of genetics in our society is a crucial factor, but the image of the detective, constructing from infinitesimally small, unconsciously shed traces the initial scenario of a crime and the unique identity of its perpetrator, also has a fascinating history of its own. The idea of scientifically tracking human traces has been attractive since the nineteenth century, and the development of this notion has contributed significantly to the power ascribed to genetic fingerprinting today. In this essay we would like to offer a series of reflections and speculations about the broad cultural assumptions which influence public understanding of techniques for tracking human beings. Our own detective work, then, will consist less of a fine-grained history of the techniques themselves than of an attempt to reveal common themes and assumptions which have been part of the compulsive storytelling generated by successive identification techniques as they have changed and replaced one another since the late nineteenth century.

Until a hundred years ago the only widespread 'trace' identifying an individual was the signature. But in the nineteenth century there began to flourish a wide spectrum of systems of human classification, based on the notion that science could produce from a peripheral trace something unique and essential about a single individual. These involved an assumption which has come to be common and extremely powerful in this century: that individuals are constantly shedding clues that can be linked to them and their actions alone; that we move, snail-like, through the world, leaving an endless array of trails behind us, trails which are infinitely small but accessible to scientific experts of all kinds. As a 1938 pamphlet to the British police, *Scientific Aids to Criminal Investigations*, explained, 'It is not an exaggeration to say that every contact must leave a trace, sometimes readily visible, but at other times only discernible to a practised eye or discoverable by an expert.'

In the half-century between the 1780s and the 1830s there emerged various systems of mapping human attributes from externally accessible signs. In physiognomy, the science that attributed psychological qualities to facial structure, one could read a person's nature in the curves of the face. Phrenology, which became popular in Europe in the same period, taught that the essential human characteristics that made up an individual could be read from the contours of the skull. Sciences such as phrenology and physiognomy partitioned individuals into an assemblage of broad types, so that a phrenologist could purportedly discern from studying someone's skull his or her moral and intellectual characteristics. According to the influential phrenologist Johann Caspar Spurzheim, the faculty of Destructiveness was roughly at the top and side of the ear. However, phrenologists could not use these physical attributes to identify a unique individual. Between the 1860s and the 1890s scientists and administrators increasingly emphasized the possibility of identifying individual people rather than type-classifying them. The confidence that these techniques could be made into systems of identification capable of indexing an entire population was a powerful lure to a society that was, as Thomas Richards has argued in *The Imperial Archive* (1993), captivated by fictional representations of an empire held together more by information than by physical force.

A number of identification techniques developed towards the end of the century promised to fulfil this goal. They claimed to distinguish individuals and provide positive and unique identities for each one. These new systems marked a shift away from conventions of human classification and interpretation embodied by sciences like phrenology and physiognomy. One such technique was anthropometry. In the 1870s Alphonse Bertillon developed a system in which length of head, width of head, length of left foot, length of middle finger and other attributes provided in condensed numerical form, on a single card, crucial physical attributes of each individual. Bertillon's system was taken up by the British Home Office in 1894 as a system of identification for convicted criminals. A second system was fingerprinting: in 1895, Francis Galton published a programmatic work on fingerprints, *Fingerprint Directories*, which asserted the individuality and permanence of fingerprints and devised a comprehensive method of classification. This method eventually superseded Bertillon's system, and following the judgement of the Belper Committee in 1901, the Home Office replaced anthropometry with fingerprinting as a technique for the identification of criminals. The actual process of taking someone's prints required less instrumentation and training and the prints promised more power of positive identification. Furthermore, an individual's fingerprints did not change during the course of a lifetime.

By the first years of the twentieth century fingerprints had become established as the official means of identification in several countries, including Argentina, the United States and England. It appeared to be a striking innovation in identifying individuals, as the case of the two Will Wests illustrated. In 1903 a witness to a murder identified a man called Will West. Police found that there was a criminal record for a 'Will West' in Leavenworth Prison, and the anthropometric details of this individual matched those of the suspect. However, the man they arrested claimed never to have been incarcerated, and when Will West's criminal history was consulted, the fingerprints of the suspect did not match. It emerged that there were two Will Wests who looked extremely similar and had matching anthropometric measurements. The story has figured for the last century in representations of fingerprinting as a powerful identification technique, because the fingerprints, uniquely, allowed police to discriminate between the two men. Because the Will Wests were black, there were racial issues implicit in the use of fingerprinting in this case, issues which we will take up later in this paper. The champions of fingerprinting claimed that this technique alone allowed experts to access the secret individuality beneath the most extensive superficial similarities.

From the late nineteenth century to the present, fingerprints and other similar traces have attracted the artist, scientist and detective of human identification. Such infinitesimal traces – ridges on a human's fingertips – literally lay on the body. And these same ridges left impressions on other objects. These traces, once subjected to the 'gaze' of science, could match individuals to a place without the testimony and aid of an eyewitness. Where physical attributes matched, as in the case of twins and similarly built individuals, fingerprints could tell people apart. In Mark Twain's novel *Pudd'n-head Wilson* (1894), fingerprints studied by a handy if sometimes bumbling detective appear to unmask people and reveal who and what they really are. Wilson, the detective, discovers the murderer of a village judge by consulting his collection of prints. Ironically, the story itself belies Wilson's own admitted shortcomings and those he professed to find in forensic science. Wilson is all too human as a detective, and hangs the following words of wisdom in his calendar:

The clearest and most perfect circumstantial evidence is likely to be at fault, after all, and therefore ought to be received with great caution. Take the case of a pencil sharpened by any woman: if you have witnesses you will find that she did it with a knife; but if you take simply the aspect of the pencil you will say she did it with her teeth.

But for everyone in the novel – Wilson, the town and the jury in the murder trial – the fingerprints alone, without any corroboration from witnesses, leave no room for doubt. The prints on the knife handle kindle Wilson's initial interest, and in the prolonged and unrealistic courtroom scenes (where Wilson appears both as a lawyer and as his own witness), Twain did not focus on legal reasoning and facts. Instead he dramatizes the way fingerprints convert jury and townsfolk to Wilson's way of thinking. The prints themselves evoke and arouse the public imagination, and the story just falls into place.

This notion that fingerprints could produce certainty *by themselves* reflected more general attitudes, even in the earliest period of their use. Even before collections of fingerprints grew to considerable size, detectives argued from theory that the odds of different individuals' fingerprints matching by chance were incredibly small – in the order of one in 64 billion according to Francis Galton's calculations. The probability of a match between two separate individuals rested on inductive arguments from small numbers of samples and on theoretical assumptions about the structure of a fingerprint. The probabilistic figure was credible not only because of the collected evidence but also because of the allure of something so unimportant carrying such distinguishing information.

With the shift from anthropometry to fingerprints came a shift in notions of human evidence. Anthropometry construed identity as emanating from the whole. That is, anthropometrics represented the *whole* individual with a series of crucial measurements that could serve as the recipe for reconstructing the physical individual. Fingerprinting, by contrast, was supposed to allow absolute identification of someone from a part that had no role in the overall physical character of the individual it identified: one's fingerprints were irrelevant to one's physical appearance or intelligence. Their virtue lay in their uniqueness, rather than in substantive links to their owner's physical or psychical character. In *Myths, Emblems, Clues* (1990) and *The Sign of Three* (1983), Carlo Ginzburg and Umberto Eco have argued that fingerprints were only the most prominent of a cluster of techniques which emerged in this period, and which were strikingly similar in unexpected ways. Each of these techniques was authoritative as evidence partly because it was unconsciously produced, and partly because, superficially at least, it looked peripheral to the crime, individual or other subject of scientific study. Individuals who worked with such evidence tracked and interpreted traces as clues to tell their tales, mimicking a host of celebrated detective figures like Sherlock Holmes in the stories of Arthur Conan Doyle, who used physical traces left unintentionally at the scene of a crime to identify the perpetrator. Less famously, the art historian Giovanni Morelli and his followers established the now standard 'Morelli method' of confirming the identity of the painter of an individual painting. This method involved examining, not the overall look and style of the painting itself, but characteristic brush-strokes unique to individual painters. And, finally, the immensely influential hermeneutic system of psychoanalysis was created during these years by Sigmund Freud, who defined a host of unconsciously produced communications – dreams, jokes and verbal slips – as being the tracemarks of profound, hidden thoughts and desires.

A crucial characteristic of each of these different kinds of traces was the fact that they were created unconsciously. The red mud on the boot of the culprit in a Conan Doyle story was left by oversight. The pattern left by the moving hairs of a paintbrush and the telling

'Freudian slip' were revealing *precisely* because they were irrelevant to what was intended to be communicated by the painter or speaker. The brush pattern was incidental to the overall image being created; the speaker's unintentional slip of the tongue detracted from the purported message for obvious reasons. However, both revealed a more fundamental characteristic of the person who made them, whether of identity or of thoughts and feelings.

Throughout the late nineteenth and early twentieth century people were fascinated by the connection between unconscious acts or traces and powerful detective work. In Wilkie Collins's *The Moonstone* (1868), the theft of a valuable jewel is carried out by an innocent protagonist while in a state of somnambulism induced by laudanum. He is not even aware after the theft that he is the perpetrator. More generally throughout the novel the notion of the involuntary trace is powerfully evoked and the link between the revealing of matters of fact and such traces stressed:

His face frightened me. I saw a look in his eyes which was a look of horror. He snatched the boot out of my hand, and set it in a footmark on the sand, bearing south from us as we stood, and pointing straight towards the rocky ledge called the South Spit. The mark was not yet blurred out by the rain – and the girl's boot fitted it to a hair.

An example – one of many – of how these conventions persisted is the Alfred Hitchcock film *Spellbound* (1945), a murder mystery based on the relations between various psychoanalysts in a psychiatric hospital. Central to the film's plot is the notion of unconsciously shed clues providing the key not only to a particular crime but to the identification of the central character, who suffers from amnesia, and his restoration to his true identity. The man arrives at a New England insane asylum announcing himself as the new head of the institution, but immediately causes suspicion by his distracted ways: he is distressed whenever he sees parallel lines of any sort. These are the psychic traces of the event on which the film pivots – a murder whose perpetrator may or may not be the amnesiac. The film parades rival modes of detection before the audience. *Spellbound* is a celebration of psychoanalysis and therefore of the possibility of scientifically constructing, from obscure psychic residues such as the amnesiac's reaction to parallel lines, an account of the real psychic reality these residues embody in compressed or disguised form. But if psychoanalysis is an example of the power of rationality to illuminate the unconscious, there is a rival 'way of knowing' in play which celebrates the converse – namely, the power of unconscious knowledge to reveal the plain facts in a case. This is the form of knowledge made possible by love and intuition. The only woman analyst at the asylum falls in love with the amnesiac and, despite her reputation for inexorable, modernist rationality (embodied by the psychoanalytic techniques she wields), it is her love that provides her with an undeliberate – unconscious – form of certainty in the amnesiac's innocence and an intuition as to the real facts obscured both by her lover's amnesia and the deceptive actions of the murderer. A combination of these two modes of psychic detection – the rational, psychoanalytical and the unconscious and intuitive – make it possible for her to reveal the murderer and absolve her lover of the crime.

The feeling that a veritable aura of unconscious, detectable traces surrounds people has been expressed again and again throughout the twentieth century, extending the belief that certain kinds of human traces, both physical and verbal clues, can be transformed into powerful narratives explaining past events and summing up the nature and character of an individual. But within these conventions 'truth' could only be found if two requirements were met. As well as the involuntary nature of the depositing of evidence, it was also necessary that the clues be tied uniquely and informatively to the individual or event. This linking could take at least two forms: either there was a unique connection between a certain clue and a certain individual or crime, as in the case of fingerprints, or the unconscious or involuntary trace held in itself some essential aspect of the individual who produced it.

For instance, consider the controversies over voiceprints in the 1960s. The commentary by Alexander Solzhenitsyn in 1968 on the new, controversial science – within a year of its invention – evoked in fiction the desperate feverishness with which voiceprint researchers were seeking to develop the technique into a new, definitive science of identity. Solzhenitsyn's *First Circle* portrays coercive, paranoid administrators in a totalitarian state searching for ever more effective means of tracking the activities of its citizens. Voiceprints seem to offer the potential to match identities with statements recorded by surveillance equipment, and a beleaguered scientist is assigned the job of developing them as a forensic tool. As the narrator complains,

This was a new science – finding a criminal by a print of his voice. Until now they had been identified by fingerprints. They called it dactyloscopy, study of the finger whorls. It had been worked out over the centuries. The new science could be called voice study – that was what Sologdin would have called it – or phonoscopy. And it all had to be created in a few days.

The extreme haste of the administrators reflects on the political system Solzhenitsyn is describing. Yet one might also argue that it reflected the allure of the notion of this new identification technique: the dual thought that voiceprints not only established certain identification, without the necessity of interposing human judgement, but, in addition, that they practically 'invented themselves' in the first place.

The underlying assumption of voiceprint analysis is that the differences between individuals in the structure of the oral and nasal cavities and the larynx, and in their use of teeth, lips and tongue can be made visible. The developers of voiceprints, like those of the polygraph, claimed that the relevant traces made by the voice are independent of conscious effort. Just as the subject of a lie-detector test is supposedly unable to control physiological indicators such as his own pulse rate, voice patterns are, according to voiceprint advocates, fundamentally 'unconscious' in similar respects. Try as one might to alter the sound of one's voice, it is impossible, according to the supporters of voiceprint analysis, to make significant alterations in the basic pattern of the voiceprint. When the human voice is recorded, these individual characteristics produce different electrical signals which influence the movement of a needle over a moving sheet of paper. The assumption necessary to the success of voiceprints is that the pattern made by the needle is not significantly influenced by deliberate changes an individual could make in the sound of his or her voice. Voiceprints, as a means of rendering speech visible, were initially developed by a trio of researchers, Peter, Kopp and Greene, in the 1940s. During the Second World War, Greene thought voiceprints could be used to identify the radio operators who accompanied troops, and therefore make it possible to track the direction and speed of their movement. Techniques were not developed sufficiently quickly to be used in this way, but they were further developed after the war, and in 1966 came the first successful use of voiceprints in a criminal trial, when an obscene caller was convicted in an Air Force court martial through voiceprint identification of the recorded calls. The scientist who conducted the examination, Lawrence Kersta, a Bell Labs researcher, testified that, in his expert opinion, the same person made both the obscene call and the known voice sample. Kersta further stated that a person's voiceprint was as unique as his fingerprint. Following this, voiceprints were used in civil and criminal trials, but they were always controversial. Solzhenitsyn's *First Circle* seemed to some scientists debating their forensic use an apt commentary on the dilemmas facing them as the technique was introduced into the courts well before they felt it had become widely accepted as a reliable identifier.

As voiceprints attempted to map individuals to vocal prints, other avenues of identification explored other aspects of the human body. Over the course of the twentieth century the traces we have developed to identify individuals have come to bear an increasingly intimate relation, at least in how we represent them, to the physiological essence of an individual. At the turn of the century, Karl Landsteiner discovered the ABO blood group system that identifies people by the antigenic properties of their blood cells; this was followed by the development of enzyme testing which allows for finer, more detailed classification. But genetic profiling is the most intimate of all, because it uses genetic material – the stuff we are taught is the ultimate blueprint from which we are made. The creator of genetic profiling, Alec Jeffreys, coined the term 'genetic fingerprinting' in 1985 to connote a crucial change in forensic science and to link his new technique with the powerful reputation of finger-printing itself. But genetic fingerprinting was not simply another form of 'fingerprinting' – it was 'a new forensic paradigm' with immense authority.

Genetic fingerprinting is on one level the epitome of what this culture of traces claims to do. In his work on clues, Carlo Ginzburg argues that what made these kinds of traces powerful was that they contained within them not only something unique to the individual they came from, but also the *essence* of that individual. In his view, the detective (whether in the form of Freud, Morelli, Holmes or the historian) performs 'minute examination of the real, however trivial, to uncover the traces of events which the observer cannot directly experience'. The implication is that one could, potentially, re-create the individual from the trace. But is this true? How would one construct the individual from a fingerprint? or the painter from the brushstroke? Such traces only seem to have the essence of their makers in retrospect, once one has the individual and the action which made the trace. This is an important qualification, because Ginzburg proposed that with these traces came a new way of under-standing and organizing society and keeping track of individuals – this innovation marked the emergence of a mode of detecting, iden-tifying and organizing identificatory traces which has dominated the twentieth century. The notion of producing the individual from the trace was central to his argument about what changed in detec-tion. And even if the nineteenth-century traces did not contain the essence of the individuals that made them, genetic fingerprinting, at last, seemed to approximate this ideal.

Genetic profiling and the adjudication of identity

Since the 1960s, genetics has proved one of the most important areas in developing new and ever more certain conventions for the scientific expert in court, and genetic evidence is regarded by lawyers and juries alike as having almost uniquely unquestionable certainty. Genetics is now both one of the most prestigious sciences of the late twentieth century and one which has immense potential relevance to people's lives. It promises to tell them who they are, where they come from and what the future (in terms of health) might bring. Genetic information, like fingerprinting of the nineteenth century, has the power of identification. This power is alluring to the state, the detective, the doctor and the single parent fighting for the paternity (or, potentially, maternity, though such cases are rare) and financial support of a child. Stored blood samples, computerized genetic databases and medical records hold biological traces or results from such traces that are immensely powerful to whomever has access and control over such items. Computers in particular allow for more control. The chance of someone slipping through the cracks is much smaller on one level, especially as under the 1994 Criminal Justice Bill in Britain police can take samples of blood from any arrested person for DNA testing, as a recent article in *Electronic World and Wireless World* (August 1994) has discussed. The power to search through endless records for a match is now mechanical, independent of exhausted detetectives and Holmeslike deductions. Individuals are merely computer records, easily differentiated, archived and searched. Ironically, genetics and genetic screening also have the effect of making us more similar in that prenatal tests may increase the number of aborted foetuses with an uncommon genetic characteristic.

Issues of state control emerge in the access to and control over genetic information. Currently, advocates of genetic profiling are pushing for record-keeping systems similar to those established for fingerprinting. The majority of states in the USA, the FBI and the British police are now establishing and maintaining databases of genetic profiles on convicted criminals and, usually, unsolved cases, leading critics to fear that aggregated information equates with control by the state. The keeping of genetic profiles is controversial: stored samples can supply not only information as to identity but also may enable medical status and other genetic traits to be determined. Moreover, the genetic profile of a convicted felon gives information on his or her biological relatives who may never have been profiled and don't want such information accessible. At the hands of the state, genetic traces are powerful checks on society. Keeping track of the traces has repercussions for political and social control. Their detection and subsequent storage makes them crucial and powerful components of a narrative of identification.

Genetic fingerprinting was introduced to the scientific community in 1985 by British geneticist Sir Alec Jeffreys and colleagues at the University of Leicester. The original project which produced the technique was an effort to find means of identifying markers for disease: images derived from a molecular pattern in the cells of the human body. The material used in the standard profiling technique consists of Restriction Fragment Length Polymorphisms (RFLPs), short sequences of base molecules, typically twenty bases long, which keep on repeating consecutively along a DNA strand at particular locations. Two factors make the pattern unique to an individual: the number of repeats at a given spot and the particular locations on the DNA molecule where these repeats are located. The method consists of extracting DNA from a specimen of blood, semen or tissue, and using restriction enzymes to cut it into fragments. These fragments are then separated by size by a process called gel electrophoresis. The DNA bits are placed in a gel through which an electrical current runs, attracting the DNA to the opposite end of the gel – the shortest ones migrate faster and farther up the gel. The DNA fragments are then transferred onto a nylon membrane, to which radioactive probes are applied to select and mark certain sequences. These tagged sequences appear once an X-ray film of the membrane is developed, and the resulting pattern of stripes on the film, looking something like a supermarket bar code, has become known as a DNA 'fingerprint'. One interesting characteristic of the genetic material involved in RFLP testing is that this material is genetic 'garbage'. It is not part of the functioning genes but just happens to be unique to each individual. This means that in its very peripheral nature, the DNA print created with the RFLP technique is strikingly similar to the fingerprint.

At first DNA fingerprinting was used to establish genetic relationships in paternity and immigration cases. But then the applications exploded into a dazzling range of uses in Britain,

Europe, America, New Zealand and Australia, and, curiously, in America they became involved with the national as well as the individual sense of self-worth and awareness in relation to crime, history and ecology. Celebrated accounts of DNA fingerprinting have helped to link it in this decade to issues and events which figure powerfully in America's self-image. DNA in the saliva used to lick an envelope linked a particular individual to the 1993 bombing of the World Trade Center; in the early 1990s University of Minnesota researchers performed DNA tests on lung tissue from a one-thousand-year-old mummified Chiribaya Indian, and found it matched tuberculosis bacterium, thus exonerating Christopher Columbus from the charge that his ships had introduced the White Plague to the new world; currently, animal geneticists are tracking DNA in endangered species to find ways to strengthen breeds; and at the Armed Forces Institute of Pathology in Gaithersburg, Maryland, DNA has identified remains of at least fifteen soldiers registered as Missing in Action whose bodies were returned to the United States two decades after they died in Vietnam.

As this litany of applications suggests, DNA profiling has been rigorously linked to some of the most powerful touchstones of American national identity, rescuing the American public from some of its best-publicized fears and guilts, and helping to fulfil some of those few agendas that all Americans can regard as a worthy cause.

'The blood will speak louder than all the rumours'? The construction of expertise and identity in 'the case of the century'

During the trial of O. J. Simpson following the double murders of Nicole Brown Simpson and Ronald G. Goldman on 12 June 1994 the American public was bombarded by various narratives of the murder constructed from the material and verbal evidence, and by a weight of evidence so voluminous and so saleable that it was compiled on CD-ROM and sold on the high street. At the centre of this chaotic picture was the DNA profiling procedure. The physical evidence included blood-streaked Reebok sneakers, a glove, and socks

located on O. J. Simpson's estate; flesh removed from under Nicole
Brown Simpson's fingernails, and bloodstains found at the Brent-
wood apartment, adjacent sidewalks and in O. J. Simpson's white
Ford Bronco. These were sent to the labs with samples of Simpson's
and the victims' blood to be profiled. The press made DNA fingerprinting all-decisive for a trial which they represented as an all-con-
suming feature of American life. Despite the technical nature of the
evidence, the details were explained – and flaunted – as yet more
alluring aspects of an addictive case. A typical headline from *New
York Newsday* on 24 June 1994 proclaimed DNA 'the test that could
nail or clear O J'. Columnist Jim Dwyer promised that 'the blood
will speak louder than all the rumours'. The media continuously
provided helpful explanations on the DNA testing procedure; perfum-
eries and game companies have capitalized on the publicity with
colognes called 'The Double Helix' and genetic profiling board
games; and, more generally, the press both celebrated the powers of
DNA profiling and, just as frequently, inverted the very notion of a
trial, announcing that it was DNA profiling which was 'on trial' as
much as O. J. Simpson. In an assertion of the power of public judgement, the news team of LA's CBS Channel 2 television vetted the
profiling techniques by testing them on their own bodies before the
11 p.m. newscast on 24 February 1995.

The profiling laboratory in the Simpson case proclaimed a positive
match, but the fight over the significance of this test and profiling
more generally had just begun. For the prosecution, the positive
match built a 'trail of blood' on which to base its case; despite a
host of physical evidence they had neither murder weapon nor witnesses. The defence, conversely, constructed an attack on the collection and possible contamination of samples, leading the media to
investigate the practices of the Los Angeles Police Department. To
subvert the immense power of the DNA traces the defence chose to
attack not the techniques themselves but the handling and possible
contamination of the original samples instead. It was not the
mechanized DNA procedures that were undermined but the more
vulnerable links in the chain of forensic authority, the supposedly
all-too-human police officers and forensic specialists who handled

them. In his opening statement defence lawyer Johnnie Cochran painstakingly separated the medical and forensic uses of DNA. He called the LAPD a 'cesspool of contamination' and emphatically claimed that blood specks 'are not fingerprints'. Cochran wanted to dispel the fingerprinting analogy of the DNA evidence in the minds of the jury. Now accepted as certain in the public imagination, traditional fingerprinting offers an alluring sense of absolute proof in a crime story. The Simpson defence wanted to undermine the public's awe of the newer DNA technology, to attempt to locate it in a contested space rather than an authoritative one.

The pieces of evidence were dissected physically, psychologically and legally. It is striking that the process of examining these minutiae involved the whole nation; experts separated by thousands of miles could virtually share the same courtroom and examine the same samples. For instance, a Harvard law class was connected on-line to the most up-to-date information about the legal proceedings. The electronic traces were stored and forwarded and dissected both on the screen and in class. The DNA matches spun a story in themselves; the double-helix structure of the blood specks on the sidewalk outside Nicole Brown Simpson's home promised to produce a narrative of guilt or innocence. They also, according to the defence, hid stories of possible mishandling and corruption. As the trial coverage blasted out all day and every day through most of 1995, saturating the national networks, the material evidence in the case promised to deliver a drama far surpassing the soap operas their coverage was replacing. One intimate narrative had been replaced by another. The blood-signposted path through personal histories and motivations was sold as appealing and popular public entertainment.

The power of the infinitesimal and the reproduction of identity

One of the central issues in the Simpson case was, of course, that of race. Throughout the investigation the prosecution had to struggle against constant defence accusations of a racist justice system in which the colour of Simpson's skin made him more likely, in the eyes of the prosecution, to have committed the crime of which he was accused. Within this context genetic profiling tended to be used as a way of abstracting the process of criminal investigation from the embodied prejudices of real human beings. The various mechanical techniques which transformed the forensic traces in the Simpson case to the positive match had no social agendas or prejudices of their own. The fact that here, at least, one could imagine the evidence being treated in isolation of issues of racial prejudice conferred upon the technique an immense weight of trust and expectation. And of course this view of the profiling techniques had a much more general currency beyond the scope of the Simpson trial.

The representation of profiling as a technique of scientific investigation which is truly mechanical and disembodied, which has no racial agendas or social prejudices of any kind, is both immensely powerful and somewhat deceptive. When police drive into the Chicago south side, for instance, because they have learned that a particular suspect, identified by witnesses as African American, fied in that direction, the spectre of genetic fingerprinting could be used to transform the detective exercise. The search would not be aimed at locating an 'African American' person who might suffer from the stereotypes of witnesses, but rather as a hunt for a particular chemical structure – a unique series of nucleotide base pairs. This manoeuvre might then make it possible to paper over issues of race which remain salient but which would no longer confront us explicitly. Biases are subsumed into the traces and pasted over when the traces are converted into court charts and laboratory reports. The source of a sample to be tested is no longer in the spotlight: the procedures draw the sample away from its social and political environment. By alluding to chemical structures, the police, public imagination and courtroom can step around troubling and pertinent social issues. The focus is on the trace once it has been removed from the mud, sidewalk or crime scene and the biological components of the person, also separated from the social context, who left the trace behind. The environment is discarded as unimportant.

This presentation of profiling as being racially and socially neutral is belied by the very real and pervasive issues of race which surround the application of the technique. Odds of a chance match are calculated from race-specific data-bases, which led to a powerful critique of population genetics in the Simpson trial. More generally, when prosecutors consider whether or not to employ DNA evidence as a forensic tool, questions of race, gender and age are pertinent because they are involved in the evaluation of how hard the case will be to prove. That is, if the crime is known to have been committed by a black individual against a white individual, if the perpetrator is young and unstable, or if perpetrator is known to victim, then DNA evidence is less likely to be used than if the defendant is older and socially more stable, or if the perpetrator and victim are both black or both white, or if the crime is perpetrated by a white against a black person. Thus issues of race do surround techniques like fingerprinting, but they are subtle and tricky. This has always been true of some obvious techniques – for instance nineteenth-century anthropometrics. Here, clearly, is a science in which racial assumptions were important, both in how the technique itself was conceived (because it categorized people by group in terms of their physical attributes) and in its close connections with other branches of science explicitly concerned with the management of race – such as the eugenics movement. However, issues of race also operate just as powerfully, precisely *because they are less visible*, in identificatory techniques which are ostensibly colour-blind. For instance, there was the hidden but nevertheless absolutely important assumption, in the case of the two Will Wests, that it is harder to distinguish between members of a population of black individuals than among a similar group of whites. It was this assumption that gave the case its power as a representation of what fingerprinting could do. Similar issues operate today in the perception of genetic fingerprinting, and they are being debated in various areas where genetics is regarded as having immediate social implications, not only in regard to genetic fingerprinting but also. for instance, in the debates over proposed research on the genetic basis of violence: the apparently disembodied nature of the genetic procedure and the knowledge gained gives the technique and the claims it supports a deceptively interest- or prejudice-free reputation.

The notion of a machine-like or machine-performed technique capable of reconstructing the salient characteristics of an individual brings up a slightly different issue as well, namely the relationship between identification and replication. There is an interesting relationship – one which has been implicitly alluded to over the course of this paper – between identifying something by way of understanding its essential parts and replicating something on the basis of that knowledge. Simon Schaffer has suggested the power of apparently thinking and feeling devices to deceive the intelligent public: both the artifices of Babbage's world and the DNA prints of today's seduce the imagination. Babbage's project involved the notion that one could create the human capacity for intellectual productions without the human being – and more generally that, by making an automaton, scientific skill could produce the appearance of humanity, or even the essential attributes of some aspects of humanity. The detective process, once vested in an automaton, would exclude the detective, hunched over the crime scene, bagging items and trying to reconstruct a narrative of events: machines could take the samples within the plastic bags and create a reliable narrative without the detective's physical and mental exertions.

In genetic fingerprinting there is a similarly ambiguous and striking relationship between identification – namely, knowing, deciphering, decoding the person's constitutive aspects – and actually creating or duplicating that individual. Cryptography in the case of genetics and genetic fingerprinting can be understood as two related activities. It is the means by which we can trace the structure of an individual's genetic code, but it is also the dynamic involved in the production or reproduction of a human being, that is, the replication of DNA material and the chemical deciphering of the code in tissue regeneration and embryological development. So the process by which we identify an individual falls within the same category as the process by which that individual was created. And this dynamic has been represented most evocatively in public speculations about the replicating power of profiling techniques. For instance FBI special agent Kenneth Nimmich, envisaging a time of ultimate evidentiary certainty from forensic profiling, described a future technique which could produce an actual image. That is, he anticipated something similar to a photograph or video representation of the individual whose blood had been tested. He claims, 'Three pairs [nucleotides] determine the color of hair, three pairs determine the color of eyes. It's just a matter of time until we find the right pairs to draw a physical profile of a person just from their DNA.' This then, is the true fulfilment of the potential we ascribe to genetic fingerprinting: the production of the very individual who left the trace.

The irony here, however, is that those traits which we as human beings conventionally use to identify one another are actually useless to identify people from genetic material. This is because this material is shared by so many people. And therefore it is precisely those attributes which we actually use to distinguish people which cannot be used, genetically, to distinguish them from each other. For, as we have mentioned earlier, it is genetic 'trash' (physiologically irrelevant material) that is used in traditional profiling techniques. What for the 'personal' observer counts as uniqueness is precisely what for the 'genetic' observer counts as common ground. A related issue, and one which is already a central area of controversy, is the way that extensive similarities can occur in racial groups, such as blacks, Asians or Native Americans. Such built-in matches would alter the probabilities for each profiling technique, so that one would have to factor into each test the type-classification of the individual one was testing. This means, of course, that one would need sufficient information about the individual to place him or her in a racial group, and in the future this might become problematic. There is considerable dicussion taking place at the moment as to whether – at least within American urban society – the standard racial and ethnic categories even work any more. With respect to many potential candidates for genetic profiling, one would need an intimate knowledge of the family structure and history of the individuals before one could be sure of placing them in the appropriate racial type, and of course this sort of procedure would undermine the whole notion of why genetic fingerprinting is so powerful: the wonderful anonymity and impersonal character of it.

While DNA printing appears to be an impersonal, technological and computerized procedure, it tells incredibly intimate stories of the people and events drawn in and from the biological sample. DNA typing encompasses more than making or rejecting a match; these matches are stories of family relationships, medical futures, sex and murder. There is an extreme intimacy about the stories told in conjunction with genetic fingerprinting – the narratives of rape, of ambiguous parentage, of marital infidelity which often prove to be one of the central dramas of the lives of those investigated – and something significant about the fact that these narratives are validated via consultation with the genetic material which is popularly referred to as the 'book of life'. One factor, as discussed at the beginning of this paper, is what Marina Benjamin has described as 'the power of the small', the extremely compelling idea of an image being able to magnify and capture the essence of the minuscule. The Victorian microphotograph could cram a huge amount of information – a narrative of deception – into a purportedly scientific image. This might suggest that, along with the conventions of detection that were constructed during this period and become so powerful over the course of the twentieth century, the power of the infinitesimal similarly came to be entrenched in our attitudes to the study of living things.

Finally, we would like to return, briefly, to the story with which this series of reflections began: the story of *The Blooding*, in which the rape and murder of two girls famously was solved by genetic fingerprinting. Wambaugh's true story recounts how the double murder of two young women remained unsolved for years. Then, when genetic fingerprinting became available, police proposed to profile genetically all the inhabitants of the village of Littlethorpe, feeling confident that the perpetrator was in the village community. Of course no one could be compelled to undergo the tests – this would be a violation of civil rights – but the assumption was that everyone would wish to do so in the hope of revealing the murderer. One by one, hundreds of villagers were ruled out. It then emerged that one young man, Colin Pitchfork, had asked a friend to give a blood sample in Pitchfork's home. Pitchfork was arrested, whereupon he confessed to the rapes and murders, and then after his confession his blood was tested.

The story of the testing – the blooding – of this small village is an appropriate place to end because it encapsulates, in miniature, some of the most enduring issues surrounding the management of human traces in the late twentieth century. Here, in microcosm, are both the prospect of state control over a community through complete access to the 'book of life' of each constituent member, and the moral, conscientious aspirations which may lead a disparate society willingly to abdicate its power and privacy to such an all-seeing régime. In short, the 'blooding' of Littlethorpe might itself be considered a trace. Apparently peripheral, it nonetheless yields the essential insight we need into a phenomenon which is sure to prove both increasingly consequential and increasingly controversial.

Unstable Regions: Poetry and Science

LAVINIA GREENLAW

Sooner or later writers turn to what they grew up with, drawing image and metaphor from their auto-mythologized childhood. I grew up with science, in particular medicine, and my childhood, like that of someone brought up in a religiously observant household, holds many memories of ritual and mystique. My parents both trained as doctors and my father was a GP who ran a dispensing practice, so we were exposed to the sights and (quite literally) the tastes of medical life from an early age. We were guinea pigs for the mixtures he prepared, being asked to test the sweetening effects of peppermint oil or synthetic banana. Meals could be dominated by lively debates about tapeworm or yeast infections, leaving me with a complete insensitivity as to what constitutes inappropriate table talk. We were offhand about the reflected glory of being related to the person who answered the call at the cinema, service station or ice rink for a 'doctor in the house'. At school, friends would ask my advice on their injuries and I developed a convincing bedside manner. There were anatomical charts in the garden shed, ampoules of heroin in a locked cupboard, and books about sexual technique on the study shelf.

Like the children of actors and police officers, we were expected to follow in our parents' footsteps although they themselves applied no pressure. My siblings trained in chemistry, engineering and physics but I knew I wanted to write from an early age and wasted no time in rejecting the sciences at school, an easy matter in an education system that insisted on an irrevocable choice between the 'two cultures' being made at the age of thirteen. As far as I was concerned, the sciences were badly taught and had a chronic image problem.

At sixteen I was reading Shakespeare, Beckett and Eliot, and was preoccupied with infinity and death. This, coupled with the usual

teenage penchant for alienation and futility, led to a fascination
with time and space. I was intrigued by the insignificant nature of
the human scale in relation to 'reality' – the universal scheme of
things; and also by the way in which the human compulsion to
map and make sense was countered by the subconscious which
could leak determinedly, even fatally, through this measured and
measurable existence – Hamlet's dilemma: 'I could be bounded in a
nutshell, and count myself a king of infinite space; were it not that I
have bad dreams.'

Outer space rather than life on the street was the real world I
was struggling to comprehend, and science was a way to find out
more about the phenomena that intrigued and disturbed me so
much. In the years since then, I still have not properly grasped the
concepts of relativity or the space–time continuum. When I write
about them it is with one eye to the comedy of this struggle to
comprehend and its risk of dissipation:

> A phone call after the pub had shut.
> *If you free yourself of the human scale,*
> *there is only futility. If it is*
> *futility, you cannot be free*
> *of the human scale.* Like the answer
> you gave me once in a dream:
>
> *I can't see anything in this mist.*
> *Then open your eyes!* A burst of peace.
> You raced home to write till morning.
> The computer belonged to someone else.
> At five a.m., you pressed the wrong button.
> Nobody had stopped to take notes.

> from 'The Cost of Getting Lost in Space'

Maybe I did suffer from the romantic notion that the meaning of
life was hidden in black holes, that in this secular age the hidden
mysteries of outer space could offer spiritual solace. Maybe I still do.
(It is true that looking at the planets makes me feel better about
death.) But with my minimal scientific education and lack of math-

ematical thinking, if I have been looking for any meaningful answer, I have not got very far. In any case, what comfort can be found in a realm where matter itself slips through the astrophysicist's fingers? I remember talking to my brother when he was finishing his doctorate, a study of the spectra of a particular cluster of stars. He said it had been a good month, he had got a result: 'Twenty million light years plus or minus twenty million light years.'

> I meet my brother in a bar
> and he shows me a piece of outer space:
> six degrees by six degrees,
> a fragment stuffed with galaxies.
> He explains how you get pairs of stars
> that pull each other into orbit,
> for ever unable to touch or part.
> When he's gone I remember
> you, eighteen and speechless,
> and how in the attic of your parents' house
> you would take off my clothes,
> run a finger as light as a scalpel
> across my stomach, then do nothing.
> Years later, I wake in the night
> still framing the words. And I like
> the idea of those stars.

'Years Later'

Poets do not claim to be writing scientific papers in scientific language – to them, science is material not craft. So why do scientists get so irritated by what they see as poetic trespass? Perhaps it is because poets inevitably colour their ideas, turning a clear window into stained glass with all its added symbolic and spiritual associations. This process can result in glorious misinterpretation but can also be surprisingly fruitful. Poetic freedoms can be used to express scientific ideas in a more broadly comprehensible manner and can locate them within the cultural, historical and ultimately human context by which science itself is driven and defined. The pleasure and challenge of taking on such subjects comes not least from the ways in which they themselves are bound up with the limitations and variables of perception. Just as ancient philosophers drew analogies with the familiar – such as Empedocles proposing the sea to be the sweat of the earth – the human experience of science remains a human one.

I would argue that poetry and science are part of the same map but would agree that they are different countries. What they share is an economy based on perception and articulation, whether in the form of conceit or hypothesis, metaphor or proof. The borders are clear but the current excitement and interest in common territory comes in part from affinities of motive, process and preoccupation that seem, at the moment, to be particularly keenly felt. There is renewed interest in the relationship between the two disciplines, a relationship that has always been influenced by their opposing or complementary cultural positions. At times the work of scientists has appeared to be a threat to cultural vitality, at other times scientists are allies in a shared experience of discovery, question and change. Poets have portrayed science as a dehumanizing and alienating influence with enormous powers to save or destroy the world and scary implications for social control. They have also shared in the excitement of scientific achievement and celebrated the common creativity of the two fields. There have been times and places where the two were considered parts of one broader field of human inquiry, with a relationship that was both natural and unquestioned.

John Polkinghorne, a particle physicist and Anglican minister, writes of the relationship between the scientific and religious views of the world: 'Do we have to choose between them or are they, instead, complementary understandings which together give us a fuller picture than either on their own would provide?' This is not to urge nostalgia for Keats's 'haunted forest', or to suggest that the Enlightenment was a bad idea and that we should allow our perceptions to be dominated by superstition or pious sensibility. Just that a piece that does not fit the empirical picture may serve to enlarge it.

In his essay on the poet and immunologist Miroslav Holub in *The Government of the Tongue* (1988) Seamus Heaney reminds us of (in particular English) poetry's essential romanticism: 'The dream's the thing, not the diagnosis.' He praises Holub's successful balancing of intelligence with compassion, irony with delight, and his driving concern that 'the true shape of things should be common knowledge.' But what if a precise expression of the true shape of things includes the obscure or fantastic? Heaney's own poem 'Fosterling' warns against the ways in which sense can limit the senses:

> Heaviness of being. And poetry
> Sluggish in the doldrums of what happens.
> Me waiting until I was nearly fifty
> To credit marvels. Like the tree-clock of tin cans
> The tinkers made. So long for air to brighten,
> Time to be dazzled and the heart to lighten.

Poetry is a juggling act of sense and the senses: Holub at once acknowledges its incandescent nature while warning the poet not to be blinded by the light. His poem 'Yoga' (*Vanishing Lung Syndrome*, 1990) begins 'All poetry is about / five hundred degrees centigrade' but ends 'And yet, only a bad yogi / burns his feet / on hot coals.'

Robert Musil, a novelist who trained as a physicist, suggests in *The Man Without Qualities* (1930) that 'if light, warmth, power, enjoyment and comfort are mankind's primordial dreams, then modern research is not only science but magic . . .' How can poets *or* scientists resist such enchantment? Musil is also quick to point out that the realization of such dreams can lead to disappointment:

the Seven League Boots were more beautiful than a motor car, Dwarf-King Laurin's realm more beautiful than a railway tunnel . . . to have eaten of one's mother's heart and so to understand the language of birds more beautiful than an animal psychologist's study of the expressive values in bird-song. We have gained in terms of reality and lost in terms of the dream.

The actual may be a rude awakening from the dream of the possible but the connections between the two remain unbroken within the

broader context of our experience of the world. We board trains
and planes but somewhere in our hearts still travel on horseback.
Poetry builds on this and makes further connections, and can at its
best explore resonance and implication without embroidering
meaning.

Heisenberg's uncertainty principle has been fundamental in the
shift towards a new scientific paradigm which gives importance to
the observer's involvement with the observed. He also asserted that
'Even in science the object of research is no longer nature itself, but
man's investigation of nature.' This is the new science, that Holub
describes as 'struggling with the "fluent" nature of things', where
discoveries such as Heisenberg's in fields like quantum mechanics
have expanded the limits of human perception, still more those of
articulation. Science is now in the exciting position of influencing
philosophical thought. Language and logic are fighting to keep up.

The mechanics of perception have always been a major poetic
theme. In an early letter, Elizabeth Bishop quotes the philologist M.
W. Croll on the baroque prose style which conveys 'not a thought
but a mind thinking . . . an idea separated from the act of experiencing it is not the idea that was experienced. The ardor of its conception in the mind is a necessary part of its truth.' Bishop's exactness
and continual and patient acknowledgement of this process characterize her work.

> The monument is one-third set against
> a sea; two-thirds against a sky.
> The view is geared
> (that is the view's perspective)
> so low there is no 'far away',
> and we are far away within the view.

> 'The Monument' from *Complete Poems*

A more difficult consideration is the legitimacy of perception that
is not first-hand and this, again, is a point on which scientists dispute the poetic exploration of their world. I have written about
science extensively but without conscious intent and always with

trepidation. After a while I have come to recognize certain dangers and have become suspicious of how such poetry can be the product of brief and casual research trips into a subject seductively rich in dazzling and unusual metaphors, images and turns of phrase. I am wary, too, of scientific anecdotes in which the words carry such resonance they can stand as their own superficial meaning. Having finished one book in which science was a dominant theme, I began to write local and simple poems, veering away from special effects. Science still fascinates me, and it is a world I will continue to explore, but I believe it shares the romantic allure of poetry's more traditional subjects, and big words and beautiful images should have to work harder than others to earn their place. I want something to be more than just intriguing or unusual, and want to see something that, following Eliot's proposition that 'Bad poets borrow, good poets steal', the poet has not simply rearranged but has made their own.

Gilbert White, the pioneering eighteenth-century natural historian, spent almost his entire life in Selborne, a parish of thirteen square miles that he studied minutely. He held very much to the ideal that one should study a subject directly and not through the media of pictures and books. Henry Thoreau, the poet who made a study similar to White's of Walden Pond, was of like mind. He urged that

a transient acquaintance with any phenomenon is not sufficient to make it completely the subject of your muse. You must be so conversant with it as to *remember* it and be reminded of it long afterward, while it lies remotely far and elysian in the horizon, approachable only by the imagination.

Compare this to a world in which we see so much further so much faster but almost always through a page or screen. How many times have you heard a poet introduce a poem with, 'I heard this on the radio/television ... read it in the paper ... saw a photograph ...'

If so much of a poet's information and experience, and therefore their data, comes second-hand in this way, how can they make it their own? Perhaps, after all, in the way that Thoreau describes, through memory and imagination. This is the power of the imagination anchored in true comprehension, as Wallace Stevens asserts: 'In poetry at least the imagination must not detach itself from reality.'

There is also significance in what we choose to observe, as choices are by definition exclusive. Each line of inquiry involves a series of decisions between a number of paths or possibilities; so the aperture closes. Koestler suggests in *The Sleepwalkers* that moments of great discovery are frequently matched by an inability to perceive other things – as if the focus must be drawn in sharply by a surrounding blackout. Of Galileo, Ptolemy and Aristotle, he says that 'It looks as if, while part of their spirit was asking for more light, another part had been crying out for more darkness.' What is seen is totally bound up with ways of seeing, to the extent that the universe is formally defined as the total of all matter, energy and space that we are *capable of experiencing* or that we have grounds for postulating as being out there.

As with issues of perception, historical relativism is a preoccupation that contemporary scientists and poets seem to share. There is a common desire to create fixed pictures, even if fixed just for a moment, out of matter at once recognized to be both unstable and to a large extent unknown.

> Crossing the Channel at midnight in winter,
> coastline develops as distance grows,
> then simplifies to shadow, under-exposed.
>
> Points of light – quayside, harbour wall,
> the edge of the city –
> sink as the surface of the night fills in.
>
> Beyond the boat, the only interruption
> is the choppy grey-white we leave behind us,
> gone almost before it is gone from sight.
>
> What cannot be pictured is the depth
> with which the water moves against itself,
> in such abstraction the eye can find

no break, direction or point of focus.
Clearer, and more possible than this,
is the circular horizon.

Sea and sky meet in suspension,
gradual familiar textures of black;
eel-skin, marble, smoke, oil –

made separate and apparent by the light
that pours from the sun on to the moon,
the constant white on which these unfixable

layers of darkness thicken and fade.
We are close to land, filtering through
shipping lanes and marker buoys

towards port and its addition of colour.
There is a slight realignment of the planets.
Day breaks at no particular moment.

'Night Photograph'

The recognition of the present caught up not only in its past but
also in its future is not a new one – 'Since Ptolemy was once mistaken over his basic tenets, would it not be foolish to trust what
moderns are saying now?' (Montaigne, 1580). But it does seem to
be a prevailing characteristic of both recent poetry and scientific
debate. The critic Helen Vendler offers a succinct definition when
writing of Robert Lowell that his later work had 'largely given up
one of the most solacing aspects of conventional lyric – the transcending of past and future in favour of an intense present moment
... Lowell forces us to read the emotions of the moment in the
knowledge of emotions past and anticipated.'

Poets are interested in how things work – nothing excites me
more than things making extraordinary sense. For me, excitement
and beauty lie in the connections and processes that cause phenomena as much as in the phenomena themselves. When I finally
took on one of the biggest poetic clichés of them all, the sunset, it
was the effects of atmospheric pollution that prompted me to do so.

I drive back along the river
like I always do – not noticing.
Then something in the light tears open
the smoke from the power-station chimney,

each twist and fold, the construction
of its slow muscular eventual rise
and there, right at the edge of it,
a continual breaking up into sky.

And that's another thing, the sky.
How the mist captures what's left of sunset,
the sodium orange and granite pink
distilled from scattered blue.

from 'From Scattered Blue'

With the domestication of technology, is it not inevitable that
scientific imagery and vocabulary become a more integrated part of
cultural expression? While few people can explain the exact workings of a fax machine, microwave or computer, such accessories are
an increasing part of even the poet's everyday life. Computing termi-
nology, in particular that of the information superhighway with its
associated fictional cyberpunk world, is entering everyday language
in the same way that Americanisms have seeded themselves all
over the world through cultural and commercial means.

The suggestion that science is coming in from the cold is increased by the recent surge of the scientist as personality. Scientists
have become prone to a little trespass themselves, using the tools
of the artist to tell their story. They appear as storytellers, even
conjurors, in their desire (from whatever motives) to gain public
attention, if not understanding. We are dazzled with images beamed
from probes on suicide missions into deep space, or the extraordinary 3D enlargements of minutiae revealed by a scanning
electron microscope. To convey ideas but also to captivate their
audience, scientists have heightened not only their language but
also the colour of their pictures (in the case of image enhancement,
literally so).

Our experience of AIDS, BSE, ozone depletion and one natural and technological disaster after another has given us a more cynical view of scientific infallibility and control. But we still look for general certainties and when scientists began to share the born-again fervour of the new physics, Chaos Theory and the Theory of Everything with the general public, it is hardly surprising that the public latched on. There is evidently widespread annoyance among scientists at the lay appropriation of such glamorous concepts. It seems that despite the protestations of Stephen Hawking *et al* that there is no spiritual significance in all this, people have clearly found it. For instance, to the unscientific eye, Chaos Theory has a liberation from logic that is dizzyingly seductive. It is in fact pure maths. The appearance of a group of controversial 'corn circles' in the shape of Mandelbrot's Set (used to map unstable regions in Chaos Theory) epitomized the way in which such notions have been imbued with New Age mystique.

Scientific language is full of unscientific resonances, derived as it can be from mythology or simple misunderstanding. In his essay 'The Language of Chemists I', Primo Levi gives an insight into the haphazard etymology of names such as benzine, originally a resin packaged by Arab merchants as 'Java incense', or *Luban Giavi*, a name that lost its first syllable in Italy and France to become *Bengiavi* and then *benzoino*. Levi is also frank about the scientific ego: 'In running through a list of names of minerals one is confronted by an orgy of personalities.' While it is true that scientific technique is not personality driven – the scientist cannot select the laws he or she must observe – the contention that one of the key differences between scientists and artists is that the former keep themselves anonymously buttoned up inside their white coats has always been dubious. Levi also discusses formulaic language, and how structural formulae articulate characteristics that are otherwise inexpressible. This search for expression is similar to the nuance and counterpoint that charge a successful poetic line. On the other hand, both poet and scientist have not only to compensate for the limitations of language but also to contend with its inherent and uncontrollable reverberations:

> People stop me in the street, badger me
> in the check-out queue
>   and ask 'What is this, this that is so small
> and so very smooth
>   but whose mass is greater than the ringed planet?'
> It's just words
>   I assure them. But they will not have it.
>
>   Simon Armitage, 'Zoom'

There is a relationship between poetry and science. It is shaped by culture and history, by affinities of subject and motivation, and the influence of paradigm and epistemology on perception and response. Inspiration, discovery and connection are integral to both disciplines, as are tedious hours spent learning a craft and practising technique. Poetry still suffers from the preconception that it is wilfully obscure. Science is an equally unpopular topic of conversation but its validity is at least agreed upon. Poetry pushes language beyond its everyday parameters to capture and express complex truths. Science has a different use of language as well as a language of its own. The two fields may not talk much to each other but right now they seem to be talking more about each other. Academics may sniff at the popular science titles decked out with gimmicky chapter headings, flashy anecdotes and brilliant full-colour illustrations, but who cares if it helps those of us brought up with equations on a blackboard to get a better grasp? Does poetry not do the same thing – the rearrangement of fact and fiction in order to convey something more clearly?

Forty years ago, science was insular, exclusive and quietly confident. British poetry was insular, exclusive and resolutely shy. The perception of both has changed remarkably in those few years and right now the borders are cautiously open, with both sides developing a healthy export–import business, while also being given to occasional friendly skirmishes. The relationship will no doubt change again. As Hawking says, in a firmly scientific context, 'Only time (whatever that may be) will tell.'

Tom Paine and the Internet

JON KATZ

If any father has been forsaken by his children, it is Thomas Paine. Statues of the man should greet incoming journalism students; his words should be chiselled above newsroom doors and taped to laptops, guiding the communications media through their many travails, controversies and challenges. Not so. A fuzzy historical figure of the 1700s, Paine is remembered mostly for one or two sparkling patriotic quotes – 'These are the times that try men's souls' – but little else. Yet Thomas Paine, Professional Revolutionary, was one of the first to use media as a powerful weapon against an entrenched array of monarchies, feudal lords, dictators and repressive social structures. He invented political journalism, creating almost by himself a mass reading public aware for the first time of its right to read controversial opinions and to participate in politics.

Between his birth in 1737 and his death in 1809, enormous political upheavals turned the western world upside down – and Paine was in the middle of the biggest. His writings put his own life at risk in every country he ever lived – in America for rebellion, in England for sedition and in France for his insistence on a merciful and democratic revolution. At the end of his life, he was shunned by the country he helped create, reviled as an infidel, forced to beg friends for money, denied the right to vote, refused burial in a Quaker cemetery. His grave was desecrated. His remains were stolen.

A popular old nursery rhyme about Paine could as easily be sung today:

> Poor Tom Paine! there he lies:
> Nobody laughs and nobody cries.
> Where he has gone or how he fares,
> Nobody knows and nobody cares.

Certainly that's true of today's media. The modern-day press has become thoroughly disconnected from this brilliant, lonely, socially awkward ancestor who pioneered the concept of the uncensored flow of ideas and developed a new kind of communication – journalism – in the service of the then radical proposition that people should control their own lives.

In the USA his memory has been tended in the main by a few determined academics and historians, and a stubborn little historical society in New Rochelle, NY, where he spent most of his final, impoverished days. In Britain the Thomas Paine Society (president, Michael Foot, MP) is widely regarded as no more than a harmless hobby for old lefties. But if journalism and the rest of the country have forgotten Paine, why should we remember another of history's lost souls?

Because Paine is for the taking and he is worth having. If the old media – newspapers, magazines, radio, and television – have abandoned their father, the new media – computers, cable and the Internet – can and should adopt him. If the press has lost contact with its own spiritual and ideological roots, the new media culture can claim him as its own. For Paine does have a descendant, a place where his values prosper and are validated millions of times a day: the Internet. There, his ideas about communications, media ethics, the universal connections between people, the free flow of honest opinion are all relevant again, visible every time one modem shakes hands with another.

The Net offers what Paine and his revolutionary colleagues hoped for in their own new media – a vast, diverse, passionate, global means of transmitting ideas and opening minds. That was part of the political transformation he envisioned when he wrote: 'We have it in our power to begin the world over again.' Through media, he believed, 'we see with other eyes; we hear with other ears; and think with other thoughts, than those we formerly used.'

Tom Paine's ideas, the example he set of free expression, the sacrifices he made to preserve the integrity of his work, are being resuscitated by means that hadn't existed or been imagined in his day – via the blinking cursors, clacking keyboards, hissing modems, bits and databytes of another revolution, the digital one. If Paine's vision was aborted by the new technologies of the last century, newer technology has brought his vision full circle. If his values no longer have much relevance for conventional journalism, they fit the Net like a glove.

Paine's life and the birth of American media prove that information media were never meant to be just another industry. The press had a familiar and profoundly inspiring moral mission when it was conceived: information wants to be free. Media existed to spread ideas, to allow fearless argument, to challenge and question authority, to set a common social agenda. Asked about the reasons for new media, Paine would have answered in a flash: to advance human rights, spread democracy, ease suffering, pester government. Modern journalists would have a much rougher time with the question. There is no longer much widespread consensus, among practitioners or consumers, about journalism's practices and goals.

Of course, the ferociously spirited press of the late 1700s that Paine helped invent was a very different institution, dominated by individuals expressing their opinions. The idea that ordinary citizens with no special resources, expertise or political power – like Paine himself – could sound off, reach wide audiences, even touch off revolutions, was brand-new to the world. But Paine could not have foreseen how fragile and easily overwhelmed these values and forms of expression would be when they collided with free-market economics. The rotary press and other printing technologies that made it possible for him to broadcast his pamphlets also led newspaper publishers to make papers tamer and more moderate so that their numerous new customers wouldn't be offended. Paine once warned a Philadelphia newspaper editor about the distinction between editorial power and the freedom of the press. It was a caution neither the editor nor his increasingly wealthy and powerful successors took to heart: 'If the freedom of the press is to be determined by the judgment of the printer of a newspaper in preference to that of the people, who when they read will judge for themselves, the freedom is on a very sandy foundation.'

So it is. But if the old media has blithely ignored Paine, it is incumbent upon the new to pay heed. The digital age is young, ascending, diverse, and already nearly as arrogant and, in parts, as greedy as the mass media it is supplanting. It faces enormous danger from government, from corporations which also control much of the traditional media, from commercialization and from its own chaotic growth.

Paine is a guide, the conscience that can help new media remember the past in order not to repeat it. He can cut across time.

Paine often introduced his most controversial ideas formally and courteously: 'The following notion is put under your protection. You will do us the justice to remember that he who denies the right of every man or woman to his own opinion makes a slave of him or herself, because they preclude the right of changing their own minds.'

This notion is put under your protection, too: Thomas Paine is the illegitimate father of the Internet. Thomas Paine should be our hero.

The sad part of Paine's story is that it is necessary to pause here and tell it to those who may have never heard any of it. He lived a life that would make the cheesiest Hollywood screenwriter blush. Born in England in 1737, he ran away from home to sail as a privateer, then worked as a staymaker and matched wits with smugglers as a customs collector. He quit to lobby Parliament for better pay for himself and fellow customs collectors. He lost his job but met Benjamin Franklin, who urged him to move to America and with whom he maintained a lifelong correspondence.

One of the regulars at Independence Hall, Paine was a philosophical soulmate of Thomas Jefferson. He fought and froze with his buddy George Washington at Valley Forge. George III badly wanted to hang Paine because he helped touch off the American Revolution with his writings, then tried him for sedition after Paine had the gall to return to England and lobby for an end to the monarchy.

He fled to France, where the bloodthirstier leaders of the Revolution ordered him killed because he urged leniency for the members

of the overthrown régime and because they feared he would alert
Americans to their increasingly undemocratic uprising. Clergymen
all over the world cursed him for his heretical religious views. Businessmen despised him even more for his radical views about labour.

In between was high drama, great daring, narrow scrapes – wan-
dering Revolutionary War battlefields dodging British bullets, fleeing
England twenty minutes ahead of warrants ordering his arrest,
coming within hours of being guillotined in Paris. Paine seemed to
live most happily in boiling hot water.

The Big Concept man of his time, he advanced deep ideas that still
resonate. An end to monarchies and dictatorships. American independence from England, of course. International federations to promote development and maintain peace. Rights and protections for
labourers. An end to slavery. Equal rights for women. Redistribution
of land. Opposition to organized religion as a cruel illusion or a
corrupt hoax. Public education, public employment, assistance for
the poor, pensions for the elderly. And, above all, a fearless press
that told the truth, gave voice to individual citizens, tolerated opposing points of view, transcended provincialism and was accessible to
the poor as well as the rich.

He was as astonishingly productive as he must have been obnox-
ious, mouthing off about everything from yellow fever to iron-bridge
construction. Although he wrote countless articles and pamphlets,
his core works are four powerful, sometimes beautifully written,
flaming-with-indignation essays.

Common Sense was the argument for independence which helped
spark the American Revolution. *The Rights of Man*, an essay written
in support of the French Revolution, attacked hereditary monarchies
and called for universal democracy and human rights. *The Age of
Reason* challenged the logic behind organized religion's grip on
much of the western world and *Agrarian Justice* called for radical
reforms in the world economy, especially in land ownership. The
first three constituted the three bestselling works of the eighteenth
century.

In 1774 Paine was thirty-seven and arrived in Philadelphia with
little more than a letter of reference from Franklin. He landed a job

as executive editor of a new publication called *Pennsylvania Maga-zine*. In January of 1776 *Common Sense* went on sale for two shillings.

Common Sense became America's first bestseller, with more than 120,000 copies sold in its first three months, and possibly as many as half a million its first year – this in a country whose population was three million. Newspapers, then crammed with controversial viewpoints, scrambled to reprint it. It had, wrote a contemporary historian, 'produced most astonishing effects; and been received with vast applause, read by almost every American; and recom-mended as a work replete with truth.' It was nicely written too, one of the first and most dramatic of the anthems and calls to arms that run through Paine's writing.

The cause of America, wrote Paine, was the cause of all mankind.

O! ye that love mankind! Ye that dare oppose not only the tyranny but the tyrant, stand forth! Every spot of the old world is overrun with oppression. Freedom hath been hunted round the globe. Asia and Africa have long expelled her. Europe regards her like a stranger, and England hath given her warning to depart. O! receive the fugitive, and prepare in time an asylum for mankind.

Paine's democratic republicanism had deep British roots. He might have been influenced by some of the world's earliest, least-known and best political journalists such as the late seventeenth-century pamphleteers Sir William Molesworth and Walter Moyle. Such highbrow English republicans had no notions of democracy or universal suffrage – not to mention representative government, which they considered anarchic and dangerous. Those were Paine's additions. He broadened his definitions of 'the people' to include labourers, slaves, women, fishermen and artisans. Paine's writings about these new notions of democracy in *Common Sense*, wrote Jeffer-son, 'considered useless almost everything written before on the structure of government.'

The publication of *The Rights of Man* made Paine the most contro-versial figure in Britain. He was followed by government spies, targeted for propaganda attacks and gossip campaigns. He experienced more fear than ever before, and for the first time was threatened with loss of freedom of speech. No book had ever sold like it in Britain. By May of 1792, 50,000 copies had been sold. *The Rights Of Man* broke every existing publishing record. There was only one possible outcome to Paine's challenge to the monarchy and the government of William Pitt, and his defence of the revolutionary stirrings in France.

On 21 May, the government issued a summons ordering Paine to appear in court on charges of seditious libel, which carried a possible – and probable – death penalty. At the end of the first week of September, his trial looming, Paine's friend the poet William Blake warned him not to go to his house or he would be killed. On 13 September he fled to Dover, boarding a ship to Calais and France while an angry dockside crowd jeered.

Three months later, the Honourable Spencer Perceval rose in Guildhall to denounce Paine as a traitor and a drunk. Defenders of Paine's republican notions failed to impress the jury. Mr Campbell the jury foreman explained that he had been instructed by his brother jurors to save time by avoiding deliberations and delivering an immediate verdict – guilty. Thousands of Paine supporters had gathered outside, shouting chants of 'Paine for ever!' and 'Paine and the liberty of the press!'

Herewith, to be put under your protection, some of the more striking connections between the Net and its spiritual father. Paine would have loved the Internet's inclusiveness. For Paine, moving ideas from one place to another at all was a spiritual notion, a miraculous vision. He imagined a global means of communication, one in which the boundaries between sender and receiver were cleared away. The freedom to send and receive these ideas was, to Paine, one of the fundamental rights of mankind. And it was the essence of media. He shared this notion most intensely with his soulmate Thomas Jefferson. The two corresponded constantly about how ideas were conceived and moved.

Paine called for a 'universal society', one whose citizens

transcended their narrow interests and considered humankind as one entity. 'My country is the world,' he wrote. The Internet has, in fact, redefined citizenship as well as communications. It is the first worldwide medium in which people can communicate so directly, so quickly, so personally and so reliably. Where computers are plentiful, digital communications are nearly uncensorable.

This reality gives our moral and media guardians fits; they still tend to portray the computer culture as an out-of-control menace harbouring perverts, hackers, pornographers and thieves. But Paine would have known better. The political, economic and social implications of an interconnected global medium are enormous, making plausible Paine's belief in the 'universal citizen'.

He would recognize the Net's style and language too. Paine believed that media should speak in short, spare, unadorned language that everyone could understand. His writing brims with humour, sarcasm, exaggeration and paradox. He was the first modern political writer, writes John Keane in *Tom Paine: A Political Life*, the newest and perhaps the best of the Paine biographies, 'to experiment with the art of writing democratically and for democratic ends. He hammered out a colloquial style that eschewed meaningless sentences, purple passages and general humbug because he considered the highest duty of political writers was to irritate their country's government.'

Reading Paine is eerie after spending time online in online political conferences on the WELL, say, or poring through the most provocative e-mail. From 'smiley' emoticons to reasoned arguments to raging flames to the staccato shorthand (LOL, IMHO) of countless e-mailers, digital communications are spare, blunt, economic and efficient. Paine's style is the style of the Internet; his voice and language could slip comfortably into its debates and discussions.

If Paine would feel at home there, he would also fight to protect this nascent medium. Learning what had happened to the media he founded as corporations moved in, he would spot commercialization as Danger Number One. He believed in a press that was not monopolistic but filled, as it was in his time, with individual voices, one that was cheap, accessible, fiercely outspoken. He believed that media like the Net – many citizens talking to many other citizens – were essential to free government.

He was right: journalism's exclusion of outside voices and fear of publishing any but moderate opinions has made it difficult for the country to come to grips with some of its most sensitive issues – race, gender, violence. Media overwhelmed and monopolized by large corporations, inaccessible to individual people and motivated primarily by profit was the literal antithesis of Paine's life, his work and his vision for the press.

We could use his clear direction at a time when mainstream media are losing their ethical grounding. Paine would never appear on talk shows or garner fat speaking fees. At one point during the Revolutionary War – when he was completely broke, as usual – he was offered a thousand pounds a year by the French government to write and publish articles in support of the Franco-American alliance against Britain. Paine said no. He told friends that the principle at stake – the freedom of political writers to express opinions free of any party's or government's taint – was sacred, even if it meant being a pauper. And for him, it did.

During his life his value system remained intact. Shortly before he died, bedridden, impoverished and mostly alone, he fired off a note to an editor in New York City who had messed with the outspoken prose in one of Paine's final essays. 'I sir,' Paine wrote, 'never permit anyone to alter anything that I write; you have spoiled the whole sense that it was meant to convey on the subject.' His deathbed scene was perhaps the greatest example of Paine's refusal to compromise.

Lapsing into unconsciousness, in agony from gangrenous bedsores, Paine woke occasionally to cry, 'Oh, Lord help me! Oh, Lord help me!' Convinced that Paine's time on earth was nearly up, a physician and pastor named Manley took advantage of one of Paine's last lucid moments to slowly save his soul by saying, 'Allow me to ask again, Do you believe, or let me qualify the question, Do you wish to believe that Jesus Christ is the son of God?' Incapable of acquiescence, even when it might have provided him some comfort,

Paine uttered his quiet last words: 'I have no wish to believe on that subject.'

Small wonder one colonial wrote of him:

The name is enough. Every person has ideas of him. Some respect his genius and dread the man. Some reverence his political, while they hate his religious, opinions. Some love the man, but not his private manners. Indeed he has done nothing which has not extremes in it. He never appears but we love and hate him. He is as great a paradox as ever appeared in human nature.

It is easy to imagine Paine as a citizen of the new culture, issuing fervent harangues from *http://www.commonsense.com*. He would be a cyber hellraiser, a Net fiend.

He might belong to contentious conferencing systems like the WELL or New York's Echo, but he would especially love cruising the more populist big boards – Prodigy, CompuServe, AOL. He would check into Time On-Line's message boards and tear into Republicans and Democrats daily. He would e-mail the *New England Journal of Medicine* his tracts on the spread of disease, and pepper *Scientific American* home page with his ideas about bridges.

He would bombard Congress and the White House Internet site with proposals, reforms and legislative initiatives, tackling the most explosive subjects head-on, enraging – at one time or another – everybody. The Net would help enormously in his various campaigns, allowing him to call up research papers, download his latest tract, fire off hundreds of angry posts and receive hundreds of replies. They would hear from him soon enough in China and Iran, Croatia and Rwanda. He would not be happy to find his old nemeses from the House of Hanover still around in Britain, but would be relieved to see George's heirs reduced to tabloid fodder and France a republic after all. He would emit Nuclear Flames from time to time, their recipients emerging singed and sooty. He would not use smileys. He would be flamed incessantly in turn.

He and the massing corporate entities drooling over the Net would be instantly and ferociously at war as he recognized Time-Warner, TCI, the Baby Bells and Viacom as different incarnations of the same elements that scarfed up the press and homogenized it. The gap between Paine's tradition and modern journalism seems poignant and stark. Journalism no longer seems to function as a community. Since it no longer shares a value system – a sense of outsiderness, a commitment to truth-telling, an inspiring ethical structure – journalists seem increasingly disconnected from one another as well as from the public.

Paine would have lots to say about the so-called Information Highway and the government's alleged role in shaping it. One of his pamphlets – this may be the only thing he would have in common with Newt Gingrich – would surely propose means of getting more computers and moderns into the hands of people who can't afford them.

He would be spared the excruciating loneliness he faced in later life on that modest farm, where neighbours shunned him and visitors rarely came and where he pored over newspapers for any news of his former friends' lives. No longer an outcast, thanks to the Net, he would find at least as many kindred spirits as adversaries; his cyber-mailbox would be eternally full.

Instead of dying alone and in agony, Paine would spend his last days sending poignant e-mail all over the world from his deathbed via his Powerbook, arranging for his digital wake. He'd call for more humane treatment for the dying. He would expound online about the shortcomings of medicine and the mystical experience of ageing while digging into his inexhaustible supply of prescriptions for the incalculable injustices that still afflict the world.

'I know not whether any Man in the World has had more influence on its inhabitants or affairs for the last thirty years than Tom Paine,' John Adams wrote to a friend after Paine's death in 1809, 'for such a mongrel between Pigg and Puppy, begotten by a wild Boar on a Bitch Wolf, never before in any Age of the World was suffered by the Poltroonery of mankind, to run through such a career of mischief. Call it then the Age of Paine.'

It is odd that so spectacular a force of media and political nature should be so vaguely remembered. Unfortunately for Paine, the historian Crane Brinton reminds us, revolutionaries need to die young or turn conservative in order not to lose favour with society. Paine did neither and fell from grace. Many of his reform programmes will always remain unacceptable to resurgent political conservatives; his religious views will always offend Christians. Though his memory is invoked from time to time, 'his resurrection will never be complete'.

At the moment, though, he is showing signs of minor respectability. This year, officials in Washington were considering funding a monument to him somewhere. And Sir Richard Attenborough, the famed actor and director, has been struggling for several years to get studio backing for a film about Paine. A Paine bio – which at the very least would feature two bloody revolutions, stand-offs with Napoleon, tangles with the House of Hanover and cameo roles for Washington, Jefferson, Robespierre and King George – would make a socko TV mini-series, too.

Imagine the scene of his near execution. Paine went to France after the Revolutionary War as a hero and supporter of democratization there. But the French Revolution was far bloodier and more violent than America's. Paine tried to save King Louis's life and pleaded with the country's new rulers to be merciful and democratic. Eventually he was imprisoned and sentenced to death. In June 1794, six months into his harrowing imprisonment, Paine fell into feverish semiconsciousness. His cellmates barely kept him alive, mopping his brow, feeding him soup, changing his clothes.

The prison governors were to take him to the guillotine the next morning. At 6 a.m., a turnkey carrying Paine's death warrant walked quietly down the prison corridors, chalking the cell doors of the condemned, marking the number 4 on the inside of Paine's door. Usually the turnkey marked the outside of the door, but Paine was seriously ill and his cellmates had been granted permission to leave the door open so that a breeze could cool Paine's profusely sweating body.

That evening, the weather cooled and Paine's cellmates asked a different turnkey for permission to close the door. Knowing that the number on the door was now inward, the occupants of the cell waited, Paine murmuring on his cot. Near midnight, the death squad slowly made its way down the corridor, keys jangling, pistols drawn. One of his friends cupped his hand over Paine's mouth. The squad paused, then moved on to the next cell.

A few days later, the Revolutionary government was overthrown. Despite his close call, Paine stayed in France until 1802 when he managed, inevitably, to alienate Napoleon. At the invitation of Jefferson he returned to a hostile welcome in the United States, where he spent the last seven unhappy years of his life.

Perhaps, if a movie is made and Paine becomes a focus of attention once more, somebody could locate his bones. That they are missing may be the most fitting postscript to his life. In his *Weekly Political Register*, William Cobbett, under the pseudonym of Peter Porcupine, smarted at the way Paine had been neglected in his final years. 'Paine lies in a little hole under the grass and weeds of an obscure farm in America. There, however, he shall not lie, unnoticed, much longer. He belongs to England.'

Just before dawn one autumn night in 1818, Cobbett, his son and a friend went to Paine's New Rochelle farm – the hole under the grass is still there, marked by a plaque from the Thomas Paine Historical Society – and dug up his grave, determined that Paine should have proper burial in his native country. The story grows fuzzy from there. By most accounts, Cobbett fled with Paine's bones, but never publicly buried the remains. Some historians think he lost them overboard on the return voyage. But certain British newspapers report their being displayed in November 1819, in Liverpool.

After Cobbett's death in 1835, his son auctioned off all his worldly goods, but Paine's bones weren't among them. Parts of Paine, truly by now the 'universal citizen' he wanted to be, have been reported turning up intermittently ever since. In the 1930s, a woman in Brighton claimed to own what clearly would be the best part of Paine to have – his jawbone. As historian Moncure Daniel Conway wrote a hundred years ago: 'As to his bones, no man knows the place of their rest today. His principles rest not.'

Candied Porkers:
British Scorn of the Scientific

NEIL BELTON

English culture is full of antiquated barricades, tollgates at crossing points between forms of imagination that should be free and open. One of the strangest is the neo-Gothic door that controls access from science to other kinds of making and thinking.

Yet this must be a ridiculous argument. Britain is a powerful western country that has its science-based industries and is as dependent on technology as any other. Since the seventeenth century it has produced versions of some of the modern world's most definitive and powerful ideas: gravity, electromagnetism, atomic physics, molecular biology. When the home islands had the strongest economy in the world, the Empire seemed to celebrate its scientists and glory in its engineers and discoverers. Even Darwin was laid to rest in Westminster Abbey 'among the illustrious dead', in one newspaper's words, attended by lords, bishops and commons. The coffin of the man who had dethroned Man was carried by a couple of dukes, a dean and the Darwinist élite. 'The State must take and enshrine the body', as Adrian Desmond and James Moore write in their account of the brief, irresistible campaign to bury this free-thinker in the prime site of the state religion. Not pulling any of their social-determinist punches, they close their vivid biography – *Darwin* (1991) – with a peroration on the triumph of natural selection as political theory. They see it as political not just in slipping its leash as a theory of the mechanism of evolutionary change and being captured by a conservative establishment, but political from its birth, a rationale for social as well as natural fate:

There were new colonies, new industries, new men to run them – not least a 'new Nature', as Huxley called it, speaking through new priests, promising progress to all who obeyed. Darwin's body was enshrined to the greater glory of the new professionals who had snatched it. The burial was their apotheosis, the last rite of a rising secularity. It marked the accession to power of the traders in nature's marketplace, the scientists and their minions in politics and religion . . . for Darwin had naturalized Creation and delivered human nature and human destiny into their hands.

This makes me uneasy, first because Desmond and Moore's performance – and it is a real showstopper – seems to short-circuit the connection between Darwin's social interests and the way in which his theory functions as knowledge. The social becomes an exhaustive explanation of the scientific. 'This move', as the philosopher Alan Chalmers says about such vanishing acts in his *Science and its Fabrications* (1990), 'is much too swift.' Is Darwin's theory adequate as a story of how organisms change? Can you pitch it against reality and still find it coherent, despite its ideological origins? A century of biological research governed by Darwinism seems to say that you can.

I am even more uneasy, though, because their conclusion seems wildly optimistic. These ironists of scientific history overrate English culture's capacity to appropriate science. The symbolism of interment in the nave of the Abbey can be read in a different way. The state buried Darwin's bones all right – buried them deep. One hundred and twelve years after his death, this state could not care less about serious research. (For the purposes of this discussion, the state will defiantly be endowed with a distinct personality.) The Institute for Fiscal Studies, in a paper by Rachel Griffith and others, claims that the UK is the only Western country in which R & D has steadily declined over the past fifteen years. In homilies from the great and the good, meanwhile, the chronic national inability to use science productively is treated as though it were a psychic fault, an inferiority complex instead of a structural problem.

The speculations that follow only begin to explore these questions, but they may reflect certain aspects of the last great non-republican culture.

The imaginative blockage about science persists despite well over a century of missionary activity by the new priesthood among the people, with the British Association for the Advancement of Science, the Royal Institution and the Committee on the Public Understanding of Science running their secular retreats, novenas and Sunday schools. Darwin's own evangelist was T. H. Huxley, who called one brilliant popular lecture-series his 'lay sermons'. The new religion of naturalism and progress was not, quite obviously, some underground sect; it was often 'imperial science', the technology that would keep the Great in Britain. Yet Huxley's reception still needed to be controlled, in the decades following his death, by the heirs of those eminent pallbearers of human uniqueness in the Abbey. Sir Oliver Lodge's strangulated introductions to the Everyman editions of Huxley's books in the Edwardian period suggest at least some of these anxieties. Picking up on a phrase of Huxley's in the preface to his study of Hume, in which 'the modern spirit' is a harvester of truth that devours 'error with unquenchable fire', Lodge warned his readers that while 'the harvesting of truth is a fairly safe operation . . . the rooting up and devouring of error with unquenchable fire is a more dangerous enterprise, inasmuch as flames are apt to spread beyond our control' (Introduction to the 1906 edition of *Man's Place in Nature,* written in 1863); and four years later, in the front-matter of *Lectures and Lay Sermons,* he assured them that 'the pugnacious attitude . . . which, forty years ago, was appropriate, has become a little antique now.'

Lodge knew what he was talking about. Disenchantment with progress and with the apparent levelling tendencies of the modern economy, and with any tendency (*pace* Desmond and Moore) to put the scientist in charge of the cure of souls, would soon make Aldous the most acceptable of the Huxleys. His irrepressible triviality was no obstacle to his acceptance as a sage, a high-toned Bloomsbury Buddha. His arch patronizing of his grandfather in his essay 'T. H. Huxley as a Literary Man' (1935) showed that even in this heyday of scientific utopianism English literary culture had its filters screwed on tight. In a practical crit of quite stunning banality, Aldous commends his grandfather for his use of allusion, alliteration and caesura sentences. The old man's Christmas habit of making pigs out of orange peel for his children becomes a figure of art, eternal in a way that renders science ephemeral: 'The substance of a scientific paper is incorporated into the general stock of knowledge; but the paper itself is doomed to oblivion. Not so the pig made of orange peel ... In so far as Huxley is still alive, influential and contemporary, it is as the man of letters.' The cumulative process of scientific discovery somehow endorses this suffocating restriction of culture to arts and letters. As though T. H. Huxley's earliest and most powerful defences of Darwinian theory were contentless: candied porkers in a graceful tissue of style.

The popularity of Aldous Huxley's revulsion from a stupid and hedonistic mass society in *Brave New World* (1932) has overshadowed his later fictionalized assaults on his grandfather's deepest convictions: belief in reason, the educability of the (white) masses and the possibility of using science to improve peoples' lives – not to install a secular and obscene paradise, simply to make them more decent and tolerable. *Ape and Essence* (1949) is a tract in the form of a mock screenplay, a crude sketch in which slogans clash like the literally ignorant armies it mobilizes. It is the near future, and the 'angry apes' have taken over. In California, at any rate, for 'on the other side of the world, the black men have been working their way down the Nile and across the Mediterranean. What splendid tribal dances in the bat-infested halls of the Mother of Parliaments!' Meanwhile, a female baboon leads Faraday on a leash; and the leaders of rival, savage baboon armies each bring an Einstein to heel as they face each other across a battleground. The Einsteins are forced to release poison gas by their simian masters: 'We, who lived only for Truth,' they cry. The Narrator says: 'And that is precisely why you are dying in the murderous service of baboons.' The script adds gloatingly: 'A choking scream announces the death, by suicide, of 20th century science.'

In an earlier and more interesting essay, Huxley hints at one of the deepest sources of the peculiar English form of this disenchantment. 'The Olive Tree', collected in a book of the same name in 1936, is a lament for the ecological ruin of the southern French landscape, and an elegy for the hardy tree itelf, but in one extraordinary passage there occurs a politicization of the landscape itself, a sudden shift from geology to law:

The olive tree is, so to speak, the complement of the oak; and the bright hard-edged landscapes in which it figures are the necessary correctives of those gauzy and indeterminate lovelinesses of the English scene. Under a polished sky the olives state their aesthetic case without the qualifications of mist, of shifting lights, of atmospheric perspective, which give to English landscapes their subtle and melancholy beauty. A perfect beauty in its way; but as of all good things, one can have too much of it. The British Constitution is a most admirable invention; but it is good to come back occasionally to fixed first principles and the firm outline of syllogistic argument.

Huxley here touches the nervous system of his culture. This whimsical vision of rival constitutions, in which the written version is like a harsh terrace of degraded soil, overcultivated and drying in the Mediterranean sun, but essential for an intellectual vacation before returning to the temperate parkland of some Sussex valley, flicks us into stranger and harder territory than the one Huxley normally cares to inhabit: the battlements and portcullises of the English state loom through the heat haze of Provence.

For a national culture – and science is inescapably a part of that culture in the way it is produced, discussed and imagined – is guaranteed by a national state. And the form of that state will have profound effects on the form of its culture; it *determines* nothing, but it can literally charter some possibilities, licensing the paper dreams of a people. The English state is suffused with an ideology born in combat with the rationalist and sceptical assumptions that underpin modern science. As first the empire and then the national economy declined and weakened, the Westminster state itself, a seventeenth-century cabinet of curiosities, remained stubbornly immune to change. Its magical and archaic trappings – the whole ceremonially centralized operation summed up by the formula of the 'crown-in-parliament' – began to be revealed as the very stuff of the old castle, not its ornamental façade: splendid tribal dances in the bat-infested halls . . .

Edmund Burke is the great imaginative interpreter of this state

which Huxley could not help seeing as he gazed out over the exhausted terraces of the Continental Enlightenment. Burke's astonishingly powerful writings entered into the very constitution of English
conservatism; and therefore very powerfully into the ways in which
the constitution, that invisible, unwritten but very palpable charter
of national life, is understood and inherited. The natural party of
government can influence even the way in which we view nature.

Burke is the central figure of the English counter-Enlightenment
and of English hostility to constitutional reform, 'abstract' rights
and natural science's critique of tradition. He insisted on the connection between the claims of science and those of Enlightenment politics, and his apocalyptic suspicion of both still affects the English
cultural weather. (Of course, his use of the word 'science' is more
general than ours, including philosophical speculation and the
emerging 'human sciences'.) Many European Romantics, across a
political spectrum from De Maistre to Byron, shared his hostility to
measurement, abstraction and mechanics; but only Burke remains
the semi-official poet laureate of a functioning state. His legacy
makes it easier to deny that there is an imagination that is also
republican, that imagination can be excited by theory – that science
itself could conceivably be imaginative and creative.

In the preface to his early *A Vindication of Natural Society* (1756),
Burke is already playing for high stakes. He affirms that

a mind which has no restraint from a sense of its own weakness, of its
subordinate rank in the creation, and of the extreme danger of letting the
imagination loose upon some subjects, may very plausibly attack everything
the most excellent and venerable; that it would not be difficult to attack the
creation itself; and that if we were to examine the divine fabrics by our ideas
of reason and finesse, and to use the same method of attack by which some
men have assaulted revealed religion, we might with as good colour, and
with the same success, make the wisdom and power of God in his creation
appear to many no better than foolishness.

This great theme of intellectual restraint and prudence, which domi-
nates his work, interweaves a suspicion of conscious construction,
of rational criticism, of models and theoretical innovation, with a
reverence for the dead, for ritual, for inheritance and for 'necessary

imperfection'. The British Constitution is one of the most powerful sources of inspiration for this political aesthetic, and the state becomes a work of art that no 'science' should correct. 'The science of constructing a commonwealth, or renovating it, or reforming it, is like every other experimental science, not to be taught a priori.' Social life, the state's foundation, is too dense a medium for the light of metaphysical speculation; 'in the gross and complicated mass of human passions and concerns, the primitive rights of men undergo such a variety of refractions and reflections, that it becomes absurd to talk of them as if they continued in the simplicity of their original direction' (*Reflections on the Revolution in France*, 1790).

In his speech of 7 May 1782 on the constitution, he rounds on the moral chemists, architects and geometers who would tamper with nature's masterpiece:

Our political architects have taken a survey of the fabric of the British constitution ... neither *now*, nor at *any* time, is it prudent or safe to be meddling with the fundamental principles, and ancient tried usages of our constitution ... our representation is as nearly perfect as the necessary imperfection of human affairs and of human creatures will suffer it to be ... it is a subject of prudent and honest use and thankful enjoyment, and not of captious criticism and rash experiment.

Any attempt to grasp the nature of the state through mere perception is doomed to failure, for it is a historically developed form, not a matter of mere volumes and masses: 'a nation is ... an idea of continuity which extends in time as well as in numbers and space. And this is a choice not of one day, or one set of people ... it is a deliberate election of ages and generations ...'

As this speech rises to a climax, there is an extraordinary passage, a dense concoction of mythical reference and Oedipal denial, in which Burke associates the urge to reform with black magic and parricide:

I look with filial reverence on the constitution of my country, and never will cut it in pieces, and put it in the kettle of any magician, in order to boil it, with the puddle of their compounds, into youth and vigour. On the contrary,

I will drive away such pretenders; I will nurse its venerable age, and with lenient arts extend a parent's breath.

The image of cutting up the father is an extremely violent one in relation to the matter at hand – a modest increase in representation for the Commons. It seems to hint at the Titans' castrating attack on their father Uranus, at witchcraft, at cannibalism – the recurrent nightmares of the Age of Reason, of Goya's darkest paintings. (Panic seems to cling to references to the rebel Titans throughout the revolutionary period: Mary Shelley's fable of uncontrolled reason is subtitled 'The Modern Prometheus'.)

In the *Reflections*, Burke prefers a form of thought appropriate to 'a constitutional policy working after the pattern of nature': 'wisdom without reflection, and above it'; 'that practical wisdom which super-seded . . . theoretic science'. In tones echoed by generations of objec-tors to written bills of rights, he explains how our forebears 'preferred their positive, recorded, *hereditary* title to all which can be dear to the man and the citizen, to that vague speculative right, which exposed their sure inheritance to be scrambled for and torn to pieces by every wild, litigious spirit'. The slowly matured parkland of Aldous Huxley's reverie would be destroyed by the speculators and inventors of rights: 'Is every landmark of the country to be done away with in favour of a geometrical and arithmetical constitution?' Burke's intense aware-ness of the complexity of human societies makes the smallest measure of democratic reform an unbearable simplification, a kind of indis-criminate number crunching: 'Political reason is a computing princi-ple: adding, subtracting, multiplying, and dividing, morally, and not metaphysically or mathematically, true moral denominations.'

It is this hatred of direct light, 'this new-sprung modern light', that gives such force to Burke's most quoted sentence about the departure of the age of chivalry, for it has been succeeded not merely by demagogues but precisely by 'sophisters, economists and calcula-tors'. Science unsupported by tradition can only resort to terror: 'in the groves of *their* academy, at the end of every vista, you see noth-ing but the gallows'. In this sinister perspective, even prejudice is better than the disembowelment of natural feeling that the 'pert loquacity' of the Enlightenment encourages: '. . . we cherish them because they are prejudices, and the longer they have lasted and the more generally they have prevailed, the more we cherish them.'

The state, and the British state in particular, is not therefore some contractual matter that can be revised at will. 'It is to be looked on with other reverence . . . It is a partnership in all science; a partnership in all art; a partnership in every virtue, and in all perfection.' At the close of the twentieth century, it is hard not to agree with Burke, who would almost certainly have been surprised at the endurance of the partnership. It has proved particularly unbreakable in the case of education and of culture in the broadest sense, which rejoins our understanding of 'science' with Burke's. In that labyrinth of parks, quads and campuses where lives and biases are formed, where literary editors are hatched and the heirs of Bloomsbury exsanguinated, there is little encouragement to link imagination and the acquisition of objective knowledge about the world, little danger of an overvaluation of the discipline of scientific truth.

The culture's internal balance can be measured by considering in a little more detail the fate of T. H. Huxley. One of the most gifted and sombre polemicists in the language, he had to suffer, as we have seen, the posthumous condescension of his own grandson. Yet his essays have greater force than anything produced by Aldous because they 'wrestle with difficulty' (Burke) in such forceful plain language, and they remain intellectually challenging, even where later qualifications and discoveries have superseded them. But they are not read. Some invisible, fine-meshed net in Burke's partnership of civil society and state has kept them out of our atmosphere. That Huxley was, by our standards, a racist and sexist is no explanation for this oblivion. His prejudices were rather more advanced than those of most of his intellectual contemporaries, including many socialists: he believed in full emancipation for both women and blacks, but could not believe that either group had the innate capacity to compete with white males. (Murray and Herrnstein, the authors of *The Bell Curve* (1994), reproduce this argument more or less exactly as it applies to black people, but coarsen it and extend it to the lower orders generally.) Arnold and Carlyle are staples of literature; this Huxley is in the twilight zone of the archive.

At his best, Huxley combines lucid exposition of evolutionary theory with a melancholy materialist (and Malthusian) poetry. He would have liked Leopardi.

... so long as he is haunted by inexpungeable memories and hopeless aspirations; so long as the recognition of his intellectual limitations forces him to acknowledge his incapacity to penetrate the mystery of his existence; the prospect of attaining untroubled happiness, or of a state which can, even remotely, deserve the title of perfection, appears to me as misleading an illusion as ever was dangled before the eyes of poor humanity. And there have been many of them.

Evolution and Ethics, 1893–4

He has a conventional romantic sense of the transience of human history compared with the deep, vast spaces of geological time – but without the usual romantic consolations:

... this present state of nature, however it may seem to have gone and to go on for ever, is but a fleeting phase of her infinite variety; merely the last of the series of changes which the earth's surface has undergone in the course of the millions of years of its existence. Turn back a square foot of the thin turf, and the solid foundation of the land, exposed in cliffs of chalk five hundred feet high on the adjacent shore, yields full assurance of a time when the sea covered the site of the 'everlasting hills'.

ibid.

There is, of course, as his *Lectures and Lay Sermons* of 1870 display, a demagogy about Huxley, and about the assertion that

the man who should know the true history of the bit of chalk which every carpenter carries about in his breeches pocket ... is likely ... to have a truer, and therefore a better, conception of this wonderful universe, and of man's relation to it, than the most learned student who is deep-read in the records of humanity and ignorant of those of nature.

'On a Piece of Chalk'

The natural order would teach the carpenter and his fellow workers that 'it is better for himself, better for future generations that he should starve than steal' ('A Liberal Education'). Huxley believed until late in his life that 'natural knowledge' could lay the foundations of a new morality, that human life could be read off from nature.

But Huxley is not that easily reduced to his Malthusian pessimism. He saw a kind of cultural politics in the growth of science, and ridiculed the cults of authority, of tradition (the 'everlasting hills') and of established faith in Britain, along with the institutions that sustained them. He was, as well as being a fine prose writer, an angry non-socialist radical of a type unusual in Britain, a genuine modernizing liberal: 'The improver of natural knowledge absolutely refuses to acknowledge authority, as such. For him scepticism is the highest of duties; blind faith the one unpardonable sin . . .' ('On Improving Natural Knowledge').

In this essay he lashed the Oxbridge colleges as 'half-clerical seminaries, half-racecourses', while in 'A Liberal Education' he was barely more complimentary: 'what we fondly call our great seats of learning are simply boarding-schools for bigger boys'. Any serious work in England has not come from the universities, he believed, but from 'outside barbarians' like Darwin and Faraday. He was of course writing before the academic reform of Oxbridge, but he was utterly consistent about equality of opportunity. Unlike later social Darwinists, he saw an immense potential in ordinary men and women and refused to accept that class could be reread as biological fate. In the debate over education, his position was unequivocal:

And a few voices are lifted up in favour of the doctrine that the masses should be educated because they are men and women with unlimited capacities of being, doing and suffering, and that it is as true now, as ever it was, that the people perish for lack of knowledge.

Huxley's conventional Victorian progressivism is complicated by his increasing awareness that nature cannot provide moral or political solutions to specifically human problems. *Evolution and Ethics*, his last work, famously repudiates the 'cosmic process', the natural world, as a pattern of conduct for humanity. Civilization is the *re-*

moval of the conditions that gave rise to the intense struggle for existence. Huxley is scathing about the 'pigeon-fanciers' polity', the despotic utopia of the social Darwinists who would weed out the 'unfit', who should, if they were consistent, 'rank medicine among the black arts'. Their deliberate suppression of natural affection and sympathy removes all restraint on conduct except the calculation of self-interest. He dismisses this wild Darwinism as little more than 'reasoned savagery'. He exposes the fallacy that equates 'fittest' with 'best', whether in nature or human society, and insists that civilization means 'a checking of the cosmic process at every step and the substitution for it of another, which may be called the ethical process'.

Huxley's 'ethical process' is a kind of communitarian moral sense. People wish to avoid pain, to achieve pleasure, to assert themselves. This impulse, so powerful in the confrontation with nature, becomes destructive if it is given free play within human societies. Emotional and moral mimicry of acts which promote human solidarity, conventions of honour, shame or law, tends to check this chaotic pleasure principle. We judge our acts by 'the associations, as indissoluble as those of language . . . formed between certain acts and the feelings of approbation or disapprobation . . . We come to think in the acquired dialect of morals.' This is Adam Smith's conscience, 'the man within'; it is 'organized and personified sympathy'. Far from being a natural arena in which the fittest should be allowed to claw their way to the top, market society – and Huxley was not a critic of the market – should be organized so as to suppress the qualities most suitable for victory in the struggle with nature. 'In place of thrusting aside, or treading down, all competitors, [civilization] requires that the individual shall not merely respect, but shall help his fellows.'

It is difficult to see in Huxley that caricature of the natural sciences in which any compromise with the idea of an objectively defined knowledge of reality is tantamount to a letting loose of the police on human life, and which sees the possibility of using the accumulation of such knowledge for good as an illusory whiggish euphoria. His stoicism about human limits certainly owes much to Malthus's vision of scarcity as a simple tragic arithmetic limiting

human population and happiness. But he was also simply realistic.
Progress can't after all solve the pain of ageing and illness, and
Huxley, by the end, saw in 'exact knowledge' merely an amelioration of an essentially painful and sorrowful existence, offering 'a
larger hope of abatement' but no fantastic solution to the inevitability of death. This insistence on the contribution that science can
make to human dignity, a kind of relative happiness in the here
and now, while refusing to ignore the limits of human possibility, is
one of the most attractive features of Huxley's writing.

Huxley's invisibility today surely owes something to the fact that
he is a classic figure of the Enlightenment. He has little resonance
in a culture which values the purely literary above all else, and
apotheosizes the Romantic tradition most of all. Burke rules. It is
striking, for example, that Isaiah Berlin, one of Britain's most eloquent theorists of liberalism, uses the relativist and pluralist themes
he derives from the Romantic tradition to defend the Enlightenment
values of democracy – turning the tradition of Burke inside out.
Berlin's suspicion of universal reason leaves him defending 'liberty'
despite its unfortunate parentage.

Reading Huxley, I was struck not only by how contemporary he
seems but how unEnglish. To sharpen the point, it might be worth
looking at a figure whose work is in temper uncannily reminiscent
of Huxley's: Thomas Jefferson. Here is a younger contemporary of
Burke's, a thinker and politician whose work is almost literally constitutive of a state. The text of the Declaration of Independence,
which he wrote, is celebrated with a national holiday, its interpretation is an academic industry. The USA, unlike Britain, has a written
constitution – 'an ingeniously constructed Enlightenment machine
of "counterpoises"' as Garry Wills describes it in *Inventing America*
(1978). The Declaration has no such force; it remains an enigmatic
and visionary charter. These documents do not explain, but cannot
be ignored as a factor in explaining, why what Burke called the
'partnership' between the state and 'all science' has encouraged less
moat-like dykes between different forms of culture in America, and
why, for example, both the USA's greatest poet and its most epic

novelist of the nineteenth century were convinced democrats and obsessed with discovery, technology and understanding. Whitman and Melville were enabled to produce exhaustive national myths, which are also 'world books', because they were citizens with a different relationship to the state than that which prevailed in the England of Tennyson and Dickens.

(The Italian literary critic Franco Moretti, in his wonderful book on the European *Bildungsroman*, *The Way of the World* (1987), observes that the English culture of justice, rightly the envy of European democrats in the nineteenth century, had the paradoxical effect of infantilizing the English imagination, turning its greatest novels of youth, for example, into fairy tales of right and wrong – not only novels of childhood but also children's literature. That conception of justice was deliberately backward-looking, one of Burke's ineffable inheritances from the Revolutionaries of 1688 who had in turn simply 'restored' English liberties: the Revolution was an act of continuity; an astronomical turn rather than a new creation or rebellion. Juridically this heritage formed a stable and conformist world, which felt that it did not need universal norms and abstract rights; complex choices, ambiguities and identities could be avoided or negotiated in black and white terms.)

Garry Wills, in his archaeological work on the Declaration of Independence, has shown how Jefferson wrote it as *a scientist*: 'he had an artist's love for the aesthetics of Newtonian mechanics.' Strongly influenced by the ideas of the Scottish Enlightenment, he believed that the pursuit of happiness was as fundamental an impulse in humans as the force of gravity between objects, and he thought that happiness could be measured mathematically, like most things in the world. 'Nature, the undamaged norm, is presumed to be uniform.' Men are therefore essentially the same and equal. If we find this tendency to view the human body as a machine disturbing, Wills reminds us that 'it is hard for us, who live after romanticism's assault on the industrial revolution, to remember how liberating was the vision of a human *machinery* in the eighteenth century.' The senses were a beautiful mechanism, 'mechanic' a proud title. Wills insists that Franklin's and Jefferson's endless

'tinkering' and obsessive ingenuities were peculiarly American in so
far as the United States itself 'is so characteristic a product of the
Enlightenment'.

'Man', for Jefferson, should correspond to 'nature'. So far, so much
grist to the mill of Berlin and all critics of the Enlightenment. But as
Wills painstakingly shows, there was more to it than that. The
well-ordered machine of government was meant to inspire a sense
of public spirit, of benevolence, of the general good. Like his master
Francis Hutcheson, Jefferson believed that there were 'distributable
quanta of happiness' – in Hutcheson's legendary sentence, 'that
action is best which accomplishes the greatest happiness for the
greatest numbers'. The point was that 'happiness' of any kind on
earth had been a matter of theology, and the suggestion that whatever limited happiness could be achieved could actually be *computed*
by fallible human intelligence was a scandal. Not only that, happiness became 'the test and justification of any government'.

The foundation of rights was not property, or inheritance or pre-
scription; it was the public good. Jefferson followed the Scottish philosopher Thomas Reid in holding that there was an innate moral
sense, which was deeply democratic in that it was common to all
men, no matter what their capacities or class: it was literally
'common sense'. This once again was a kind of communitarian
morality. They also believed in the existence of an internal faculty
for which acts of benevolence were aesthetically pleasing, and
which prolonged that pleasure by repeating benevolent deeds: as
Wills defines it, 'virtue begins with a *spectator's* joy in the action of
a prior *benefactor*.' Virtue thus became social, the moral sense defining the ends of action and reasoning only the means.

None of these thinkers, as Alasdair MacIntyre points out in *A
Short History of Ethics* (1967), could explain why we should approve
of benevolence rather than self-interest, and they tended to assimilate
ethics to a rather unstable aesthetics. But the purely philosophical
weaknesses of these ideas are beside the point. Their echoes in
Huxley are obvious, and presumably derive from the same sources,
as crystallized in the Scottish Enlightenment. The Declaration is one
of the fundamental documents of a state, written in the light of

what was then understood to be 'science'. Its imagination of human possibilities is both more frightening, more rigidly mechanical and statistical, than our intangible, illegible and romanticized charter; it is also more generous and democratic, and more encouraging to the imitation of, rather than passive reverence for the rights it proposes.

I do not wish to suggest that there is some originative power in these documents, and that the USA is somehow admirable because it has a more interesting founding text. Scientific boosterism flourishes there alongside evangelical Protestantism; democratic liberties struggle with biblically authorized bigotry. The USA is a hideously divided society, lacking public institutions that could check the ferociously dynamic power of its market economy. It is also the real home of politicized Darwinism, of a radically progressive conservatism, as Hofstadter showed in *Social Darwinism in American Thought*, written between 1940 and 1942. We can see an aggressive revival of it in the 1990s in a figure like Newt Gingrich, with his equally obsessive interests in space stations and in limiting the fertility of the underclass. But the USA is none the less recognizably modern and has institutions and charters that are set up to deal with modern problems, and Britain does not. The argument here is simply that the realm of the possible in a culture is bounded among other things by states, by legal systems, by our understanding of rights, and that the mind can have strange borders. The English novel was different, as Moretti suggests, because of a specific history. English culture's attitude to science is equally sensitive to its historical conditions – and to its constitution.

There have always been, despite the strange conditions of its existence in Britain, attempts from within the profession to claim a larger space for the natural sciences in the wider culture. The proselytizing energy of the scientists themselves can fool us into thinking that there is a single international monolith called 'technology', and that our local relationship with it is a normal one. Between the wars, the self-assertion of science did indeed reach a sort of crescendo, in what turned out to be an astonishing detour.

The story of the rise and fall of communist science is well known now largely through the work of Gary Werskey in *The Visible College*, first published in 1978. The episode has left very few traces, except in the dusty catacombs of the book trade, where the novels of H. G. Wells or the plays of Bernard Shaw are the core of any heap of literary flotsam, and where the works of J. D. Bernal and J. B. S. Haldane yellow and decay as their cheap wartime paper catches the light. Like the lignin in the wood pulp, the Stalinism of these once huge bestsellers corroded them fast. The smell of rotten compromise – of 'Traitor Tech', as an American friend of mine labelled Cambridge in the heyday of paranoia about Fifth Men – wafts out of these pages: Bernal describing the effects of collectivization as 'very remote from immediate enjoyments . . . It was grim but great'; or solemnly praising the Stakhanov movement, a 'voluntary' system of back-breaking speed-up for manual workers, as an example of applied socialist science. We choke on this stuff, as we do on Wells's whiggish authoritarianism. (There was actually a great deal more to Bernal's work outside the laboratory than crude propaganda, though this is not the place to address it.) Yet in a curious way, 'Bernalism' was a very English affair, and never left Cambridge.

The prophets of science as politics – 'scientific socialism' – and of socialism as the only way to fulfil the promise of science included J. D. Bernal, J. B. S. Haldane and Joseph Needham. These men were haunted, like later generations of intellectual Marxists from élite English schools, by the absolute as well as relative decline of their country. Truly creative as scientists, at the cutting edge of population genetics, crystallography, physical chemistry and embryology, they saw waste, backwardness and stupidity all around them. Their instincts were, with the exception of Lancelot Hogben (the only one of the five leading members of Werskey's college not to accept Soviet diamat, and a strong critic of technological progress), anything but democratic. This was a mandarin science, folded into the dialectic of history, pregnant only with good – if only a system could be built that would give scientists their proper leading role in it. 'In the practice of science we have the prototype for all human common action,' wrote Bernal in the closing pages of *The Social Function of*

Science in 1939. Science is 'the chief agent of change in society; at first, unconsciously as technical change . . . and latterly, as a more conscious and direct motive for social change itself.' Elsewhere he posed the alternatives in ways that to our eyes can look merely monomaniacal: 'Either science will be stifled and the system itself go down in war and barbarism, or the system will have to be changed to let science get on with its job.'

Werskey insists that 'no left-wing movement ever became quite so obsessional about the scientific road to socialism as the one in Britain'. This is quite a large claim, if you consider the utter marginality of British communism and the centrality of science in a huge international movement that taught militants 'the general laws of motion' of all matter – natural, cultural and psychical – as revealed in Marxism–Leninism. But it is almost certainly true, for nowhere else was there a state culture so genetically resistant to the virus of reason – even though it was going through a brief spasm of attention to production and development at the time. The USA, as we have seen (and as Portia Dadley's essay on Edison demonstrates), had an early start in its career of celebrating technology and science, and the same can be said of the French state that descends from the Revolution of 1789. Britain's problem could easily be construed between the wars as a failure to embrace the most powerful and effective forms of modern thought, and science made the absolute standard of politics. This was an antic modernism, a weird ghost produced by the strangeness of the crumbling structure it inhabited.*

* In 1966 Perry Anderson, near the start of his assault on the old pile, using siege rams built by translated western Marxists, could grandly dismiss 'the rhymes of Spender and the fantasies of Bernal' in his 1965 essay 'Components of the National Culture'. This was as much as the brightest star of a new generation of élite renegades had to say about the great critic of 'anti-scientific capitalism': one of the more interesting minds of the prewar culture demoted to the same class as one of its worst poets. This relegation, and its implicit dismissal of the need for critical engagement with science, is a sign of an interesting problem that can't be even glanced at here: the identity of attitudes to technology adopted by the post-communist left and the romantic right in Britain.

By the 1920s, Britain's relative decline was obvious. Its industries were powerful, its empire still vast, but in most areas except pure research it was falling behind its rivals. Industrialization had, for the whole of Europe, meant imitating the combinations of free labour, capital, technology and markets first tried out in Derbyshire and Lancashire in the 1760s. Now, probably no other state in the developed world was less interested in science and its applications than Britain. In the 1900s, twenty years after Gilchrist-Thomas had invented a process for making 'basic' steel cheaply from low-grade ores, the British steel industry was still making five times as much of the older 'acid' steel, importing the ores and ignoring its huge supplies of native phosphoric ore. Britain in 1913 had only 9,000 university students compared to almost 60,000 in Germany (Eric Hobsbawm, *Industry and Empire*, 1969). British science was outstanding, but it was concentrated in Cambridge and London. The real failure, Hobsbawm concludes, was not so much one of research as of development. Britain lacked a dynamic and farsighted financial sector – in plain words, bankers who could think beyond the shortest stock-market horizon and invest in research. They wanted – the state and the City expected – reliable quick returns for shareholders, which were always available in currency deals, imperial investments and commodity speculation. No break from the frenetic rhythm of this foxtrot was going to be easy.

The figure of Sir Alfred Mond, founder of Imperial Chemical Industries and a tireless propagandist for the centrality of science to a modern economy, is a reminder that a different path was possible. In the twenty years from 1931, faced with protectionism and the threat of war, the state did insist that British banks should back innovation. Cars, artificial fabrics, planes, antibiotics, radar and television sets were produced in those art-deco factories that now look like industrial museums. This grudging modernizing interlude surely encouraged the scientistic populism of the time. The end of the war took the pressure off, however, and the City could be freed from its brief and unwilling dalliance with production and development. ICI remains an exception to the rule.

In these circumstances, it is not surprising that the strange

British-Marxist variant of scientific modernism could not last. It was swept away by its own fatal fascination with the Soviet dictatorship, by popular disgust with the unspeakable uses of 'science' in Nazi bio-babble, in the practices of Mengele's lab in Auschwitz and Japan's 'Unit 731' in Manchuria, and in the bombs at Hiroshima and Nagasaki. The scientist was at best Brecht's Galileo, swallowing his conscience in order to survive, at worst Doctor Strangelove. In the real rocket states he (very rarely she) was more likely to be an Andrei Sakharov, so engrossed in his work at the 'Installation' that 'the rest of the world was far, far away, somewhere beyond the two barbed wire fences' (*Memoirs*, 1992)

And Bernalism was also killed off by a temporary amelioration of the British system. A welfare state was dragged up to the gates of Windsor Great Park and ceremonially battered until it could be coupled on to the ancient state carriage. In the decades that followed, there were flurries of 'scientific' assessment of the hybrid wagon train as it lumbered downhill, the equivalent, it now seems, of making the postilions walk in order to lighten the load. Bernal seems to have thought, near the end of his life in 1964, that much had been accomplished: 'we are no longer concerned, as I was then, merely to vindicate the growth and use of science in modern civilization' (Preface to *The Social Function of Science*, 1989).

Harold Wilson's very Bernalist rhetoric promised the harnessing of 'Socialism to science and science to Socialism', but the reform of the semi-modern economy was unlikely to be successful as long as the premodern state was not seen as an intimately related problem, and democratization of the state was never on the agenda. The contemporary diaries of Tony Benn, one of the hottest technological revolutionaries among the new ministers, are much exercised by the struggle to reduce the size of the Queen's head on UK postage stamps. The next modernizing revolution in 1979, which put a former industrial chemist into power, only intensified the dominance of finance and services over the economy and of the 'crown-in-parliament' over the polity. At the time of writing, romantic nationalism is once again the culture's spontaneous response to a threatening wave of European, cosmopolitan rationalization.

England had already found its most articulate *native* voice in the field of literary criticism. Literature was elevated to be the moral arbiter of the humanities, and of human values generally. The extent to which 'Englishness' is anchored in an overvaluation of literature, protected from the corruption of other forms of thought and creativity, is a controversial academic topic. But it is difficult to deny the influence on the national culture of F. R. Leavis, of the left-wing 'culture and society' tradition which he provoked, of the endless generations of Bloomsbury, of the highly ordered networks of 'literary London'. For these versions of culture, science was, it went almost without saying, an insult to human values. The sterility of the Leavis–Snow 'debate' between 1959 and 1962 owes something to the deculturation of science that was already far gone by then: when Leavis spoke he resonated with the strongest frequencies in the atmosphere.

In all of this, the national culture had suffered a fatal weakening of its imaginative capacities; some of the creative energy of rationalism in English culture was snuffed out. The idea of two cultures now expressed a truly schizoid condition. The republic of the imagination was confined to the republic of letters, which was less a republic than a literary map of class resentments, in which the key was always low, pretension always the worst sin and the metrical elegance unobtrusive: realist scenes, proletarian escape clauses, sub-Brechtian dramas, low-life blokeishness and novels full of bromide sentences.

Given the scale of its defeat as a cultural influence and as an effective productive force in Britain, it is hard to see why science should still exercise the beadles as much as it does. Yet in the 1990s there has been a Great Fear, rumours sweeping the countryside of advancing columns of geneticists and geometers, of the incineration of hallowed traditions and the burning of stately piles. Even the English constitution has been threatened. Aldousian nightmares, updated for a robust and scarcity-obsessed age, have become almost routine. Polemics take aim at the bloodless vampires of reason, and new burial parties – of highbrow thriller writers and low-rent columnists, of cultural studies professors with Penguin Nietzsches spoiling the cut of their Armani jackets – totter up the nave to keep the lid on the old heretic's coffin. There is a sense of crisis, of invading European norms, abstract rights, American-style charters, all the legalistic rick burners of the Enlightenment loose around the walls.

Panics are rarely without some basis in reality. There are new energies in the culture, and the disintegration of the monarchy, for example, may presage a rebirth of the democratic and republican imagination. It is clear that Burkean reveries about English institutions are not remotely as appealing as they were when they were supported by a warlike, Protestant and imperial sense of nationhood backed up by commercial supremacy (see Linda Colley, *Britons*, 1992). Yet a couple of snapshots of what are in themselves surpassingly trivial moments – given the scale of Britain's end-of-century crisis – can still reveal the ghost of Burke lingering like a negative image in the background.

Brian Appleyard of the *Sunday Times* published *Understanding the Present* in 1992. It was an almost papal polemic (foreshadowing the Pope's own blast by a couple of years) against modern rationalism, the 'horrifically effective infant' of science and the decline of magic. This is cultural studies of a kind that Burke would have understood. 'Cartesian man' is left 'dreaming on the edge of oblivion', in a world mapped, measured and ordered to death. We have lost our unity with the world, we are nomads without organic connection to nature. The medieval cosmos, by contrast, was a friendly place where, Appleyard firmly believes, 'the human race was the point, the heart, the final cause of the whole system.' (Actually, according to E. O. Lovejoy, the medieval universe was indeed full, but it was also 'diabolocentric' – Hell was the dead centre of its cosmogony, much as Wapping was the heart of Thatcher's London.)

Science is accused by Appleyard of making the world what it is, responsible for every moment of existential anguish that modernity brings: he regrets that 'each man must hope and believe what he can'. In this parody of Burke, science inevitably has a politics: liberalism. Morally neutral, tolerant, it is the scientific method realized: 'neither government nor society as a whole provides moral direction or meaning'. Appleyard denounces the 'science-based liberal democracies' which tend towards a 'unity of unbelief'; history since Bacon and Galileo and Descartes has been 'a long tale of decline and defeat'. The horror of change is coupled with an obsession with the sense of self, of being important to ourselves, and rage against Copernicus, Darwin and Einstein for displacing 'us' from the centre of the universe. There is stirring talk about resistance movements, but it all ends in a complacent, Californian, neoconservative egoism: feel good about yourself and your culture. This is the philosophy of the slush pile. 'Science will be humbled, and we shall be free to celebrate our selves again.' Thus: 'We can have irreducible affections, values and convictions. At their most fundamental these need not be defended further because they will express only our kinship with our culture and that kinship will be beyond appeal.'

Burke in Malibu, Pio Nono on the Isle of Dogs: 'It is humanly impossible actually to be a liberal.' An unspoken question hovers: what is this thing called democracy, this merely theoretical equalizer and enforcer of the hated spirit of toleration? Our 'real' selves, present to our souls in our lived culture, should have no need of such illusory representation – which can never touch our deepest prejudices and convictions. This last bastard child of the 'zealotry and bigotry' of science cannot, for the moment, be tipped out of its cradle in polite society, but the hot light is in Appleyard's eye.

Much less comically preening, Kenneth Branagh is an Irish actor-manager with vast amounts of celluloid and money. His *Frankenstein* (1994) is a magnet for every paranoid anxiety about science in a culture aware as never before of its own fragility and backwardness. The title sequence is a child's encyclopaedia picture-caption: 'Alongside political and social upheavals, scientific advances that would profoundly change the lives of all ... The lust for knowledge had never been greater.'

Revolution and vivisection stalk the movie, the villains of an icebound comic. Imagination and love are set against science, and Frankenstein is a poet seduced by physiology, tragically betrayed into the wrong faculty. He denounces his professor's 'hard science', which is presented as a mad, De Maistrean banning of all original ideas. The code of British character acting, as rigid as parliamentary procedure, shapes our responses. Robert Hardy, the television Churchill, is the voice of conservative and orderly, but also risibly pompous, hostility to experiment. The failed reanimator, the grisly has-been who shows Kenneth how it can be done, is played by John Cleese, paradigm of comic disaster and hysterical incompetence. It is as though Bob Hope were to play Robert Oppenheimer in a film about the Bomb.

The film's final scene is a wonderful Gothic interment for the mad boffin after all its channel-surfing on the breakers rolling in from deeper waters. The captain of the stranded vessel reads the burial service, which is a stranger to the Book of Common Prayer since it includes a heavily intoned uncanonical verse from Ecclesiastes 1:18: 'For in much wisdom is much grief; and he that increaseth knowledge increaseth sorrow.'

The creature stands apart like a giant Gulag victim during this ceremony. When the ice starts to break up he seizes the torch from the captain and swims to the tiny floe carrying Frankenstein's pyre, which drifts away with Arthurian solemnity into the mist. The creature raises his torch in an operatic overarm smash before lowering it slowly to incinerate his maker and himself.

Mary Shelley's *Frankenstein* is a rich source of misunderstandings, its horror of Prometheanism endlessly recycled, but it is also a wonderful fable of *irresponsible* creativity – of a science that refuses to think of consequences. This dimension of the novel, of violated 'common sense', is lost in our culture's use of it. The creature is made in 'a kind of enthusiastic frenzy' with no thought for its inevitable human response to the fact of its existence, and none for the safety of Frankenstein's 'own species'. Capable of 'love and sympathy', the creature is driven out of the social cage which alone can create civilized relationships: 'everywhere I see bliss, from which I alone am irrevocably excluded. I was benevolent and good; misery made me a fiend. Make me happy, and I shall again be virtuous.' This lack of creative solidarity between the master and his monster – the revulsion of the one against 'the filthy mass that moved and talked' that the other represents – ruins both of them.

Fiction can decline beautifully to resolve problems, which is one reason why we are haunted by great bad novels like *Frankenstein*. The unassimilable creature is still out there on the ice-floe, despite his maker's refusal to grant his Adam an Eve. He is the spectre of genetic transgression and biological fantasy. Rather than endlessly reinventing the myth in the familiar terms of our anti-Enlightenment culture, however, it would be more creative to see it as part of that really urgent discussion about the assessment and understanding of science. This discussion does not have enough participants or importance in Britain. The efforts of official science to increase 'public understanding' too often assume the worst kinds of corporate division between our culture and theirs, denying any interesting exploration of imaginative boundaries and adopting a patronizing stance of bringing light to the heathens. Democratic institutions involving people in discussion of research priorities would be a far greater stimulus to scientific curiosity than injunctions to 'understand' what 'we' do. As I have argued, Britain has peculiar difficulties here, in that the absence of open, intelligible and written democratic standards not only makes society vulnerable to dramatic technological changes but encourages the most irrational attitudes to them. Haldane long ago pointed out rather blithely that there is no biological invention which has not been first seen as a perversion before becoming accepted and ritualized (*Daedalus*, 1923). But in a complex modern society the introduction of, for example, genetically altered organisms (a rather more radical step towards the manipulation of life than, say, the discovery of beer, which Haldane gave as an example of early biological invention) could arouse the most intense hostility to even the most benign interference with living cells.

Branagh's film contains a truly disturbing image of what is at stake. In a rewriting of Mary Shelley's narrative to give it more drive-in thrills, Frankenstein cannibalizes and reanimates his dead wife after she has been murdered by the creature on their wedding night. Helena Bonham Carter is presented to us ripped apart, stitched together again with borrowed limbs, her head and face botched, mutilated. For reasons quite other than those Branagh intended, this pathetic vision of laboratory brutality does indeed almost move you to tears – because it is so gross a caricature of the possibilities of science. Is this terrible mannequin the only image our culture throws up when it thinks to itself about what knowledge of the world and of living processes can offer us? Is any application of reason to the common good to be sacrificed to this recurrent horror, as though the remediable ills and avoidable suffering that science can touch are irrelevant, and must always be projected on to the torn bodies of the victims of human cruelty?

The Difference Engine and *The Difference Engine*

FRANCIS SPUFFORD

This book has been about real history: if not about the direct histories of science and technology themselves, about the indirect histories of those things in the culture – about the mental shadows they have cast and the new thoughts and experiences for which they have provided shapes in this century and the last.

At the book's starting point, though, stands a non-event, an aborted project, something which in sober fact did not happen. Despite government grants of unprecedented size, and expectant coverage in the technical press of the day, Charles Babbage did not build the Difference Engine, and he never even came close to realizing his plans for its programmable successor, the Analytical Engine. Visit the computing history section of the Science Museum in London to see the Difference Engine, as Doron Swade and his team have made it possible for you to do, and you witness a rich historical paradox. The handle cranks, and industrialized mathematics takes place before you. The gearwheels turn, sword blades clunk up and snick down, keeping each gear registering a whole, determinate number; at the back the rotors on the main columns turn circles at staggered intervals from one another, sending silvery sine waves rising the height of the engine with each step of the operation. The bronze and steel of the engine weigh around three tons, but at the same time the thing seems curiously insubstantial once you think through its presence on a reinforced upper floor of a South Kensington museum (in one of the public palaces reserved by the Victorians for just this sort of edifying spectacle – though not this exact one). It is *so* Victorian in its visual impact; almost exaggeratedly so. It looks like an elaboration, a raising to the nth degree of every Victorian device you ever saw, from a fob watch with its case open to a

steam locomotive whose jointed connecting rods delicately transform enormous linear thrust into huge rotary motion. It expands
on every specimen of purposeful nineteenth-century metal. It even
delivers a ridiculously clear lesson in Victorian industrial philosophy. It situates its operator, who need only count the number of
times the engine has been cranked to produce this or that order of
polynomials, in a perfected version of the position labour was supposed to occupy in the Victorian factory system. Here is an abstruse
operation of the calculus, previously the province of a trained mathematical mind, made over into a job of machine minding, at the
cost of perfect alienation from the task. Yet you are seeing a quintes-
sentially Victorian object on which no Victorian ever laid eyes,
including Babbage, whose craftsmen could not cut gears to the
necessary fine tolerances.

What can you call it? A fictitious antique? In one sense it is the
relic of a non-existent history, an artificially created survival from
a past in which Babbage, after all, succeeded. Babbage faced tangible
and intangible obstacles, both problems with materials and problems with ideas. Having conceived a machine some technical stages
in advance of his era's power to manipulate metal, he lacked a
whole set of supporting technologies for the Engine, which was
why he was constantly sidetracked by questions that had to be
answered first, about alloys and temperatures and brass-turning
equipment. His design was balanced, so to speak, on the single pole
of his own determination, instead of fitting comfortably atop a pyramid of assured skills. Likewise, he could not refer the intellectual
endeavour represented by the Engines to any established context of
ideas. There was the analogy (beloved of his mathematical collabora-
tor Ada Lovelace) between the Analytical Engine's proposed card
storage for information and the card hopper of a Jacquard loom,
but analogy – individual illuminating comparison, constructed for
the occasion by a specific effort of perception – was about the limit
of what the Engine could draw on from the culture. There existed
no shared vocabulary available to describe Babbage's intentions,
and make them obvious; no already existing notation, no con-
venient mathematical shorthand, no canon of procedures. The

Science Museum's department of computing history found itself in a position to remove both obstacles, supply both lacks, solve both problems. Supervising the production of several hundred identical metal gears only proved to be interestingly tricky for them. And so far as ideas went, they possessed the very significant power to name Babbage's enterprise as part of, well, the history of computers, in which his thinking made perfect, retrospective sense. They furnished him a belated context from current computer science – they could simply refer (a tiny illustration in itself) to his hardware and software difficulties. So the Engine is in effect a collaboration between times. Its functional, successful existence here and now vindicates Babbage's design, but the *design* alone. The Science Museum did not mean to test the Engine's feasibility then, only its feasibility now. They abolished, rather than re-creating, contemporary constraints. They acted on their own power to complete what had been uncompleted, to realize what had been unrealized. The object in the gallery therefore says nothing decisive about whether Babbage could have succeeded. Instead it pays a tribute to a man who has been retrospectively organized into being an ancestor; and it brings the might-have-been to the surface, all the more powerfully for not being decided one way or the other. When we look at it, this synthetic thing made from now's response to then, we are dealing with a relationship to the past, with our own ability to imagine and retrieve, and with the fluctuating process by which at different times different parts of history seem to come to prominence as we recognize unfinished business there. In fact we are seeing a peculiarly concrete, peculiarly provoking exercise of the historical imagination, something just as alive in the history of science as in the history of manners or the history of speech. And the imagination we use alike to 'enter' the past idly, and to reason about it carefully or proportionately, does not possess the history it works upon as some uniformly dead or distant object. There are constant transactions involved, like the transaction between Babbage and the Museum: exchanges and barters in a sort of imaginative economy. We nominate some times as livelier than others because, like the living present, they seem to contain unexhausted possibilities, of which our sense that at a certain point things could have gone otherwise gives a perverse sort of index.

The Difference Engine still radiates a feeling of possibility which cannot easily be accounted for. Certainly not from the functional point of view: by any modern standard of comparison, the Engine has negligible powers. Any programmable pocket calculator can be instructed to calculate the polynomials; and it can solve any other equation you program it to as well, while Babbage's metal beast is a one-task tool, hard-wired or rather hard-cogged for its single purpose. The shining, immovable architecture of the Difference Engine actually embodies its program. The thing's beauty in fact declares how thoroughly it is obsolete. But even if it were the open-ended Analytical Engine we were looking at, the same functional comparison would produce the same result. It is not what the Engine can do that is really at issue, but some quality of promise we see in it independent of its actual function. This quality, interestingly, the established and ordinary technologies of the present completely fail to produce. Neither a calculator, nor, say, the motherboard of an office PC (a piece of kit infinitely removed from Victorian tech in complexity and performance) promises anything except that it will perform the tasks we know it can. The Engine, on the other hand, obstinately promises wonders, and the domain of wonders is the domain of the unspecified, the rich-because-uncertain. In part, perhaps, this only amounts to an old delusion about complex systems. Unlike printed microcircuits, the components of the Engine fall within the range of scale at which made things can be readily admired. A cog of the Engine is accessibly fist-sized; yet there are so many of them that the eye cannot immediately tally the total, a much more aesthetically impressive bemusement when you lose your way in a object the size of a small bathroom. And the same subrational hope begins to operate that alchemists felt when they had multiplied the rules and formulae of their craft beyond the mind's capacity to hold them all at once: that the whole may magically become greater than the sum of the parts, and complexity tell on itself to secrete the philosopher's stone. But the promise we perceive has another side, tied to the development of technologies and the juncture in their development when wonder plays a proper part. It is the promise of beginnings. As a technological vista opens for the first time, before the journey up it has really begun that will prove each step to be prosaic, one by one, the possibility of the whole line of development can be felt at once. Often the anticipations turn out to be inaccurate, but that knowledge has not yet been forced upon us, likely and unlikely consequences have not yet been sifted apart. For a brief moment, for that very reason, modest results of an invention and frankly utopian results can have equal likelihood in our minds, and are rolled together intoxicatingly, almost lyrically. If the Difference Engine had been built, it would have represented that phase in the development of computers. Since it now has been, it enables us to glimpse the chance for the same information technology that we actually have as it would have appeared from the vantage point of anticipation – as a wonderful prospect, an opening box which could contain anything. The Difference Engine re-enchants computers. It doubles our vision: all the while word processors and accounting software go on being dull and familiar, yet simultaneously we sight on them as shadowy promises at the edge of plausibility. We can look forward to the present (a mental sleight rather like Flann O'Brien's 'art of predicting past events') in the spirit of the 1840s manual of technical breakthroughs, written for the use of alert artisans, which described the Analytical Engine even in the absence of a working model as bringing 'metal close to rationality', bounding on towards artificial intelligence at the slight hint from Babbage that such things could now be hoped for. Wonder includes terror as well as delight. It isn't the sleep of reason that brings forth monsters after all, but reason's first confused awakening to a line of inquiry. Mary Shelley wrote *Frankenstein* from the equivalent tangled moment at the birth of the life sciences. 'Perhaps a corpse would be reanimated; galvanism had given token of such things; perhaps the component parts of a creature might be manufactured . . .' Perhaps, perhaps. If we want to find the present counterpart to the Difference Engine, we have to look not at the achieved technologies it might have pointed towards but at technologies now passing through the same phase, where

the perhapses still cluster and the dispensation of wonder still ob-
tains. Virtual reality, say, or the Internet, though the bloom of unlimited possibility is already passing from them. The Internet begins to
shrink back from its ecstatic characterizations (a web for democracy,
a surfable sea of data, an instrument of expanded perception) and
settle down as a very large database-cum-techie-salon. The smart
idolatry is moving on. Nanotechnology, maybe? Still, whatever
should be the next locus of wonder, the Difference Engine will resem-
ble it more than it resembles any strictly comparable device for
crunching numbers. Phases of change remain curiously *in* phase
with each other, where the imagination is concerned.

But the history the Difference Engine stands for is of course an
imaginary one.

Imaginary histories offend against seriousness. They lead the
mind away, most historians have thought, from what can be studied
into what by definition cannot; from solid actuality, quite difficult
enough to document and comprehend, to a terrain of nebulous possibility governed by wishes and other illegitimate impulses. We
humans can know history in a profound sense, argued Vico, because
we made it, in however complex and collective and half-intentional
a way. What did not happen, we did not make, and cannot know
from within, as we can know (or grope towards knowing) the acts
and omissions even of the most distant and different peoples. This
leaves thinking about them as a hobby at best, a peculiarly ethereal
recreation, a confessed preference for unreality. Yet it must be admitted that the foolish question of how things might have happened is
certainly involved with, attached to, the sensible question of how in
fact things do happen: how necessity and chance, the quick fore-
ground incidents of history and the slow background shifts in the
regimen of human societies, luck and weather, material factors and
intellectual factors, the sequence of scientific discovery and the effect
of popular novels, all combine – differently, according to different
contending theories and understandings of history – to shape an
outcome and produce this event rather than that one. Only a belief
that all events are rigidly predetermined can entirely exclude the
possibility that things could have been otherwise. And most

historians are not pure determinists. So the same qualities that
make imaginary histories ridiculous have occasionally commended
them and made some people wonder if they could be useful.

For a while in the 1960s and 1970s a fashion waxed among
statistical economists for 'cliometry'. It was a beautiful and arrogant
term meaning 'history measuring', after Clio, history's muse. What
do you measure history against? It is a permanent problem in all
studies of unrepeatable phenomena (in geology and epidemiology
as much as in the history of humans) that you must make inferences
without being able to set up controlled experiments. Shifts in the
earth's crust, outbreaks of plague and industrial revolutions cannot
be run through to order in the laboratory. There are no control
groups, no test cases, only the single pattern of raw data provided
by events as they fell out. It's like being asked to come up with an
account of a whole game when you can shake the dice only once.
The cliometrists tried to get round the problem by referring the real
world to hypothetical worlds – 'counterfactuals' – where there had
been other outcomes. One study, for example, tested the effect that
railroads had had on the nineteenth-century American economy by
modelling a railway-less version of it and then comparing the two,
in the hope of being able to tell apart those developments that were,
and those that were not, the economic consequences of railway
building. Objections to cliometry came from those who pointed out
that the fundamental difficulty still applied. A statistical model of a
different history is a grossly abbreviated, simplified thing. To keep it
coherent you had to choke back its implications; you had to set
arbitrary limits on how far economic behaviour might branch away;
you might be able to cope with the consequences of there being no
railways, but you had to leave out the consequences of the consequences. Statisticians already made a beeline for any case where
economic behaviour was different enough from the norm, and
simple enough, for it to approach the conditions of a controlled
experiment. Therefore in cliometry you arrived at a model which
gave you no more purchase on the totality of the true historical
record than you could get from a look at some real deviant group
(say, the buggy-driving Amish), and which would be considerably

less rich and reliable besides. Cliometry did not die. It lost its glamour and its name, and became a method among others, to be brought into play in those circumstances when real history has accidentally produced the chance for comparison. A humbled form of cliometry happens when medical historians model the effects of the Japanese taking to a meat diet sooner, by investigating the homogenous group of Japanese emigrants to Hawaii, who munched beef decades before the first McDonald's opened on mainland Honshu.

Around the same time, however (you see that imaginary histories themselves have a history), the idea of other, differently ordered worlds came to seem philosophically fruitful. In 1973 the logician David Lewis published *Counterfactuals*, a short provoking book which claimed a semantically watertight (and very technical) argument for, as he put it, 'realism about possible worlds'. His central chapter gleefully hinted that people kept coming up to him at Oxford parties and asking what he thought he was playing at. In the face of 'incredulous stares', or more rarely questions about what he *meant* by possible worlds, he wrote,

I cannot give the kind of reply my questioner probably expects: that is, a proposal to reduce possible worlds to something else. I can only ask him to admit that he knows what sort of thing our actual world is, and then explain that other worlds are more things of *that* sort, differing not in kind but only in what goes on at them . . . It is said that realism about possible worlds is false because only our own world, and its contents, actually exist. But of course unactualised possible worlds and their unactualised inhabitants do not *actually* exist. To actually exist is to exist and to be located here at our actual world – at this world that we inhabit . . . It does not follow that realism about possible worlds is false. Realism about unactualised possibles is exactly the thesis that there are more things than actually exist.

This was not only a bravura demonstration of philosophical wit, though it is that, and curious-sounding as ever to non-philosophers who expect that discussion shall strike out from roughly agreed terms into the material of evidence and opinions. It was a deliberately extreme statement of a belief about possibilities which a much wider range of philosophers and social scientists only feel compelled to reject when the possibilities are solidified into complete possible

worlds. Those interested in how historical explanations work – how
something we recognize as a satisfactory description of cause
emerges – have been specially willing to admit hypothetical outcomes into the argument (so long as they stay hypothetical). It's
been argued, for example, that a whole class of explanations turns
on an 'implied counterfactual'. I employed one earlier in this essay.
I asked what conditions would have had to be fulfilled for Babbage
to build the Difference Engine, in the hypothetical case that he succeeded; pointed out that those conditions were not in fact fulfilled;
and gave that as my reason why the Engine was not built. But in a
general sense, everyone concerned with explaining and describing
history has a practical stake in the 'thesis that there are more things
than actually exist'. There has always had to be some conceptual
scope found for potential, though no such *thing* as a potential can
conceivably be identified in the finite world of objects and people, or
else we allow no room for change and becoming. There have always
had to be ways found for giving an account of a situation's potential
for change, and in the social sciences attention to might-bes and
might-have-beens has sometimes seemed like a promising source of
tools. It need not be 'dispiriting', Geoffrey Hawthorn observes in his
recent *Plausible Worlds*, that possibilities 'are not items . . . about
which [we] could be said to be certain, and thus to know.' 'On the
contrary. It promises that kind of understanding . . . which comes
from locating an actual in a space of possibles', siting the real
and single course of history among its attendant fuzz of counterfactuals. Hawthorn's proposal for using imaginary history aims, mod-
estly, at illuminating where we are, and how likely it is that we find
ourselves here. It does not aim, like cliometry, to make comparisons
with the real world.

All these, though, are ways of making imaginary history serve
real history. In one way or another, they raise the ghost of another
possibility in order to investigate the groundwork of the real; they
raise it in order to lay it again. They treat the possibility like one of
those incalculable quantities that can none the less be used in mathematics because it will be neatly cancelled from both sides before
the equation reaches its final form. The specific imaginary histories

they put forward have only the status of an illustration, or a model or an example. But it is noteworthy how much nearly every author of a rare, serious inquiry into counterfactuals enjoys the opportunity to play at changes. Often they embroider their exemplary counterfactual to pleasurable excess: certainly far in excess of the strict needs of an example. They invent sly details. Geoffrey Hawthorn, for instance, kicks off *Plausible Worlds* with an updated variant on an old speculation about what would have happened if Ferdinand and Isabella of Spain had failed to conquer the Muslim emirate of Granada in 1492. Because he then has his independent Granada add early factory chimneys to its skyline of mosques, shifting the whole industrial focus of Europe southwards, he can send ripples of revision through the learned literature of his own discipline which deals with industrialization. Foundation texts of the social sciences shimmer into slightly different form. Weber's *The Protestant Ethic and the Spirit of Capitalism* becomes *The Kharejite Ethic and the Spirit of Capitalism* by 'ibn-Weber', delicious to cite.

Some things are good to eat. Others fall into a useful parallel category invented by an Amazonian tribe, and are 'good to think'. That is, they feel good when you think them through, they give a pleasure like the pleasure of eating. And indeed you are consuming something in this kind of thinking. Rather than pursuing a thought to a conclusion, you are dining on its aspects, guzzling its ramifications, digesting it gourmand-style. The issues and concerns that would structure your thought if you were in search of conclusions do come in, but like a menu. When Hawthorn gives himself the incidental pleasure of an altered book title, or when David Lewis picks out for use such propositions as *If kangaroos had no tails, they would topple over*, they are switching for a moment into this other style of thinking. They taste, in passing, the pleasure that the topic of different worlds proffers, quite independently of whether it can be turned to philosophical account. Where imaginary histories are concerned, of course, utilitarian thinking counts as the rare exception. Almost all the thinking people do about them is of the edible kind. Which does not mean that it is negligible, for it can be as alert as serious thought to a difficulty or an intellectual knot, only to differ-

ent ends. It uses imaginary worlds for the sake of imagination, and
because it can draw on so much that the imagination understands
about history, it can be extremely revealing.

Words, which summon worlds, are its proper medium. If the
money could be raised, the Science Museum would like to go ahead
and build the printer Babbage envisaged for the Engine, to tabulate
the results of the Engine's calculations, stamping them – almost
incising them, in the manner of an old-fashioned railway ticket
punch – into broad fillets of pasteboard; which raises the bizarre
prospect, at least in my mind, of the Museum following through the
next stage, and then the next, and in effect launching a Victorian
Age of Information Technology in the upstairs gallery. Why stop?
Presumably they would have to when they reached the limit of
Babbage's own designs. To start improving on the Analytical Engine
would be to leave the province of museums altogether. And it will
take a sobering sum even to assemble the printer. But what it is
formidably expensive to execute in metal, words can perform at no
cost at all, except imaginative effort and historical sympathy.

Bruce Sterling and William Gibson's SF novel *The Difference
Engine*, published in 1988, contains a world of the 1850s altered by
the rampant success of Babbage's computers*. To bring about this
state of affairs, preceding history has been adjusted as far back as
1815, so that the ground can be prepared, politically and economically, for Babbage's little metal shoots to flourish. The ironmasters
and the savants have taken unequivocal control of Britain, as they

* It isn't the only recent book to seize on Babbage. The much inferior *In the
Country of the Blind* by Michael Flynn (1990) features feuding secret societies
who employ working Babbage engines to foretell the future, an esoteric science
named 'cliology' in counterpart to genuine cliometry. Nor are Gibson and
Sterling by any means the only SF authors cannibalizing Victoriana. Apart
from the 'steampunk' writers Jeter and Blaylock, there have been two notable
British books in the last year or so: Colin Greenland's *Harm's Way* (about
Victorians magically let loose on the whole solar system) and *Anti-Ice* by
Stephen Baxter (Victorians winning the Crimean War by A-bombing
Sebastopol). We are in the middle of a small wave. Perhaps, as one SF critic
suggests, there seems just now to be an irresistible ironic match between
Victorian belief in benevolent progress and 1990s scepticism.

were prevented from doing in fact by orchestrated, conserving reforms. Instead of fissuring under pressure of diverging interests, the pro-industrial coalition of capitalists and workers and scientists which in our world only preserved a frail existence during the 1820s and 1830s, overthrew the Duke of Wellington in a revolution around 1830. Now reason rules triumphant. It is the Young Men's Agnostic Association which proselytizes. Babbage and Lyell and Brunel sit as 'Merit Lords' at Westminster, influential, idolized, masters of the public purse. Data dances unimpeded across the land by telegraph, on punched cards, on spools of ticker tape. In the bowels of the monstrous Egyptianesque pyramid of the Central Statistical Office on Horseferry Road, serially connected Analytical Engines process dossiers on every man, woman and child in the country. 'How many gear yards do you spin here?' 'Yards? We measure our gearage in *miles* here . . .' Smaller Engines check customers' credit on the counters of smart emporia, automate manufacturing, perform mathematical reconstructions of dinosaur skeletons. The Irish potato famine has been avoided by rational relief measures, but London groans beneath an overload of smoke and fumes, tottering towards systemic crisis. We recognize this city. It's a Gustave Doré drawing made more so. It's Dickensian London subjected to an exponential growth in both wonder and horror. It's the *ville* of penny-dreadful night, coloured in shades of mud and India ink, stage-lit by gas for melodramatic glare, abounding in spectacular villainy, fantastically a-swarm. In broad outline it's also, since Gibson and Sterling make amazingly conscious and resonant use of genre conventions, somewhere that cool American practitioners of the 'steampunk' SF sub-genre have visited several times before. A rejigged Victorian London seems to serve (it's been observed) as a favourite kind of artificial unconscious, where anxieties can be condensed, and discharged, on the understanding that this lurid place is definitively *other*. But genre is chiefly a consideration here because science fiction, by nature, offers both surprises and guaranteed pleasures. SF, no matter how good it is or how individual, tends always to hold out more specific and more predictable promises to the reader than fiction *per se*. To a much greater extent than the literary novel, it allows you to select the pleasures you plan to have in advance. This concentration, this extra single-mindedness, is what makes it the pre-eminent form in the culture for exploring ideas that are 'good to think'. Sometimes SF does little else. Gibson and Sterling, interestingly, come down emphatically on the side of genre convention in their choice of a precise moment for their alteration of history to begin. SF 'alternative history' uses a very simple rationale: events branch at points of decision. It shows audacity to nominate some juncture as decisive that doesn't specially look it. Despite research so thorough that historians of science who read *The Difference Engine* feel they are breathing air thick with allusions to their own work, Gibson and Sterling did not opt for a metallurgical breakthrough, or a paradigm shift in mathematics. They had Lady Byron decide she would stay with her husband in 1815, however digusting his taste for practices expressible only in Latin. And the *imagination* is tickled, outraged: satisfied.

A partial bibliography of imaginary histories drawn up in 1980 lists about 175 fictions of one kind and another, not counting films, plays and boardgames. By now the number must have multiplied many times over. Alternate history has been one of the boom areas in the genre since the (relative) decline of 'hard', starships 'n' rayguns SF. Branching points have multiplied correspondingly, but the most popular probably remain the fall of the Roman Empire, Columbus's arrival in the New World, the Spanish Armada, the American Wars of Independence and Civil War (the latter much revised at every stage from Fort Sumter to Lincoln's assassination) and the Second World War. The pronounced bias towards American history of course reflects the dominantly American readership and authorship of English-language SF. Almost all these most favoured junctures for historical alteration are military events, because a war going the other way provides the most blatant and immediate possibility of change. Usually, though, a technological shift follows. It's rare for a surviving Roman Empire not to have mastered steam power and the printing press at least, and more often machine-guns and television as well. Success for the Spanish Armada on the whole signals an aborted Industrial Revolution in Europe, with the connection between Protestants and machines, Catholics and obscurantism taken as read: under the thumb of the Popes, ox carts generally lumber along rutted roads carrying cargoes of illuminated manuscripts. Similarly, unless the author happens to be a committed Dixie partisan, the result of a Confederate victory in 1862 or 1863 tends to be a backwoods, tar-paper-shack North America, stuck with clunky telegraphs and puffing billies, the internal combustion engine nowhere to be seen. German victory in the Second World War, on the other hand, can offer the prospect of accelerated but malignant technology, Gruppenführer Wolfgang Something replacing Neil Armstrong on the moon as he steps from his swastikaed V10 rocket years before 1969. It is impossible to exaggerate how routine these manoeuvres have become in science fiction; but equally important not to assume that the conservatism, the recourse to stereotype manifested in them, is purely a generic thing, sign merely of a genre *running* according to type.

With a few exceptions people first began to play at altering history in the first decades of this century. The two collections of essays most commonly quoted as the foundation stones, F. J. C. Hearnshaw's *The 'Ifs' of History* and J. C. Squire's *If It Had Happened Otherwise*, were published in 1929 and 1931. (Interestingly, many of the contributors belonged to that governing caste in British society which had reason by 1930 to wish that history were working out differently, yet could still believe, if barely, that decisive moments were formed by the decisions of people like them. One contributor to Squire's collection was Winston Churchill. Another speculated about what would have happened if the General Strike of 1926 had not been defeated by the then Home Secretary, Winston Churchill.) But some of the attractions of altering history had been identified centuries earlier, at the point when fiction of any kind, with its sober relation of non-existent events, seemed itself to hold out the problem of a counterfeit reality. Francis Bacon classed the invented people and invented emotions of poetry as a category of 'feigned history' in *The Advancement of Learning* (1605). Poetry's root in the Greek *poiesis*, 'making', was uppermost in his mind:

The use of this FAINED HISTORIE, hath beene to give some shadowe of satisfaction to the minde of Man in those points, wherein the Nature of things doth denie it, the world being in proportion inferioure to the soule . . . and therefore it was ever thought to have some participation of divinenesse, because it doth raise and erect the Minde, by submitting the shewes of things to the desires of the Mind, whereas reason doth buckle and bow the Mind unto the Nature of things.

Imaginary history – applying Bacon's terms as he certainly never intended we should – is poetical history, *made* history; and it too gives the shadowy satisfaction of raising 'the Minde' higher in importance than 'the shewes of things'.

It can be everybody's mind that imaginary history bestows this satisfaction on. Especially when a story of what might have been plays up the extent to which history is chance-made, and therefore turns on accidents and on choices as opposed to laws or necessities, it pays in one sense a warm tribute to everyone's powers of decision. 'There is no history,' Gibson and Sterling have their scientist hero Mallory shout provocatively, '– there is only contingency!' Which seems to promise that the course of events is a piano and not a pianola: we too can sit and pick out a tune that pleases us. If history is not inevitable, and does not operate by general rules, then it may really matter whether we turn right or left when we leave the house in the morning. In a much more obvious sense, and simultaneously, sovereignty over the imagined order of things belongs entirely to the mind of the author, and the reader participates far less than she or he does in real history: the reader can contribute nothing from memory or shared inheritance of the past. Imaginary history makes its author's mind the sole arbiter of chance and likelihood; it makes over the superlatively intractable stuff of what has been into material; it turns history plastic and sculptable; it seats the author in the office of time. Spare kingdoms and republics fall out of the author's pockets like loose change.

Neither the sculptural satisfaction nor the privileged authorial position are entirely denied in the writing of real history. Inevitably narrating the real past is also a matter of interpretation and tacit modelling, and when the history in question is perceived as having a conclusion which touches on the present position of the narrator, then too the historian has a seat in the midst of history; it flows towards her or him. Michelet, for example, included his own birth in his annals of the French Revolution, where it was indeed a significant event, because the Revolution led to the possibility of Michelet becoming what he was. The rise of 'the people' from subjection to power was also, as Michelet saw it, his own rise from poverty to learning. The Revolution created the conditions for a republican intelligence like himself to grasp the times, and tell them. Imaginary history exaggerates, and literalizes, this role. The tacit becomes the explicit. A maker of imagined history has all the powers of the real historian to impose a narrative on events, tamping this and that into place in an intelligible design, with the additional power of adding or inventing the pieces to be tamped.

The heady ability to create worlds brings in its train the equally heady chance of criticizing them, as you can only criticize *made* things. When the fabric of history becomes a chosen arrangement, it can be savoured like a picture (or like the book it actually is), assessed for consistency, admired in some parts but regretted in others. An imaginary history is open to criticism from base to crown. In John Crowley's *Great Work of Time*, an exquisite and paradoxical fable about the defects of granted wishes, the 'President *pro tem*' of a society dedicated to the preservation of the British Empire by altering history wanders the streets of a place his society has, in a strange, half-accidental way, created. This 'capital city of an aged empire' answers his desires uncannily, though he never anticipated a class structure contrived from different species: saurian butlers, maned magi as subtle and gentlemanly as Sherlock Holmes, tubby hominids to do the heavy lifting, and sexless angels annunciating in the public parks. But do the pieces fit? the President wonders. He studies this world as if it were a set of nested fictions of differing, and perhaps incompatible, moods and styles:

The lives of the races constituted different universes of meaning, different constructions of reality; it was as though four or five different novels, novels of different kinds by different and differently limited writers, were to become interpenetrated and conflated: inside a gigantic Russian thing a stark and violent *policier*, and inside that something Dickensian, full of plot, humours, and eccentricity. Such an interlacing of mutually exclusive universes might be comical, like a sketch in *Punch*; it might be tragic too. And it might be neither: it might simply be what is, the given against which all airy imaginings must finally be measured: reality.

The sting in the tail, as the President finds, is that inconsistency has no force as an objection against what actually is, if with sudden doubt and bewilderment you should find yourself in the posture of a connoisseur before that real order of things which cannot be accepted or rejected, appreciated or deprecated. What Crowley makes happen to the President *pro tem* within the kaleidoscopic precincts of *Great Work of Time* can also happen in our solid and consensual world to the reader of imaginary histories. They can have a temporarily estranging effect on your perception. As you close some story in which matters are otherwise, you can catch yourself regarding the real course of history, by a kind of persistence of vision, in the guise of an equally gratuitous arrangement. If imaginary histories have any radical quality at all, it's this momentary gift of the feeling that real history might well be done better. It takes mountainous effrontery, though, to mark reality low for inventiveness and coherence. For that reason Madame de Staël reserved it as a role for the devil, God's would-be competitor at world-creating, who, in Goethe's *Faust*, she said, 'criticizes the universe like a bad book'.

But the pleasures of upset, inversion and irony are quite as attractive in the writing of imaginary histories as the chance at omnipotence. If they are a game, they are one with peculiarly loose rules. Probability is a consideration, but the decisive factor is what you can get away with, while retaining a sufficient *atmosphere* of probability. Strictly speaking, in *The Difference Engine* Gibson and Sterling cheat. They have invented a world of sluggish mainframe computers, analogous to the IBM-dominated globe of say 1960, yet wanted to infuse it with the cyberish attitudes of the present, and people it at least in part with the denizens of the hacker subculture of the 1980s and 1990s. They want youths in frockcoats who feel the same covetous delight about what they can do with a pillar of gears that youths in baseballs caps feel about the havoc they can wreak with a modem. Hence the constant tug they exert on the Babbage technology towards devices ever further out of developmental sequence. Their cinema-screen-sized 'kinotrope', with its tens of thousands of mechanically rotated pixels, lets them have Victorian computer graphics despite the absence of the cathode-ray tube. Hence too the conceptual shove in the novel which makes 'catastrophist' interpretations of evolution shade over into chaos theory. Accelerated modernity and accelerated postmodernity are all mixed up; but then buttressed by matter-of-fact storytelling, and defended by jokes and subterfuges that deliberately blur period. Gibson and Sterling work tricksily at the reader's belief. They are both keen to have you credit their alterations (though the kinotrope may be nearly·as functionally implausible as Fred Flintstone's rock-hewn TV set, it exudes the same ingenuity as real Victorian optical toys) – and eager to trap you, if they can, into spurning unlikely but genuine Victoriana. Surely calling the radical governors of this England 'the Rads' is an obvious exercise in forged slang? It hints at radiation. It sounds off-key, an unwise attempt to make Victorian politics sound zippy and street-wise. Nope: louche politicos drawl the word in Disraeli's *Sybil* (1845), one of Gibson and Sterling's chief sources, a book whose plot, characters and even author all appear here in mangled form.

Like most of those in the world-revising business, Gibson and Sterling are fascinated by the encapsulated meanings of words and habits and objects. Really perfunctory stories in the genre will indicate a different world by looking at banknotes, coins, flags, stamps and military uniforms, the most concentrated tokens of a course of history. (Even the most economical fiction featuring an American stamp with Hitler on it can't match mathematical logic for high-handed brevity. 'Possible worlds are mentioned by means of the lower-case letters h, i, j, k,' runs one of David Lewis's footnotes; 'sets of worlds by means of capital letters; and sets of sets of worlds by means of script capitals.') Such self-contained signs, above all, are portable. They work as counters that can be moved elsewhere, still meaning what they mean, so the expectations they represent collide. Imaginary histories offer the pleasure of seeing familiar objects in

unfamiliar settings: a revolver in Julius Caesar's hands, a word processor on Disraeli's desk. And the contrary tickle of satisfaction when
unfamiliar circumstances lap a familiar scene: J. F. Kennedy trun-
dling across Dealey Plaza in a cart pulled by donkeys. Likewise imaginary history can confirm, or reverse, our sense of a historical character, either extending a person as we already imagine they were
into new but suitable contexts, or using the revisionary power of the
genre to force an incongruity. The latter ploy comes accompanied
by the cynical suggestion that character and role are themselves
accidents, and could have been otherwise. A story published around
the time of the Rolling Stones drugs trial in London played with a
counter-world where Victorian manners, social deference and cold-
bath culture still ruled. It therefore contained a fresh-faced detective
constable from Scotland Yard by the name of Michael Jagger. 'This',
as Merlin remarks bitterly in T. H. White's *The Once and Future
King*, when his familiar spirit supplies him with a sailor hat to wear,
'is an anachronism . . . That is what it is, a beastly anachronism.'
Most imaginary histories use anachronism in one way or another,
though perhaps it is too broad a term for the variety of switches
and swaps that are possible with the capsuled material of history.
When a tract of time can be made to appear, like a hat out of thin
air, in the form of some typical thing, anachronism itself can be a
form of wit; of the kind of conceit-building wit which involves finding an unlikely affinity or point of comparison. Anachronism dispenses with the time between what it brings together in the same
way that calling love a pair of compasses dispenses with every obvious dissimilarity. Your surprise is the object of the exercise. A medieval magus in kiss-me-quick headgear, the Emperor Caligula riffing
away on a cherry-red electric guitar, John Keats loading every rift
of a *graphics program* with ore: all these exemplify the quality defined
(disapprovingly) in the eighteenth century as 'the unexpected copu-
lation of ideas'. Interestingly, it was around the same time as the
rejection of conceited wit that 'anachronism' first entered the vocabulary of literary critics. A growing awareness of historical difference led to stringent displeasure at current poetry and translations
of ancient works, which broke the decorum of period – by inaccurate

dress on Roman orators, impossibly modern sentiments in the mouths of Celtic heroes. Writers of imaginary history have returned the compliment and largely ignored the eighteenth century. Hardly anybody has thought it worth altering the seventy-odd years before the French and American Revolutions. Somehow, before the revolutions brought in their unpredictable fire, the eighteenth century seems too primly stable; somehow its events seem to be of the wrong kind. Nobody has ever bothered to make the War of Jenkins' Ear come out the other way. Imaginary historians prefer times that teeter between possibilities: wigless days when it seems anything might happen. They have a vested interest in flux.

All the same they surprisingly often have the people who inhabit their different history sense that something is awry. 'It has come to me,' says a scruffy aristocratic communard in *The Difference Engine*, a boy doubly opposed to Babbage's world of capitalist merit, 'that some dire violence has been done to the true and natural course of historical development.' On the words 'true' and 'natural' red flags signalling dangerous irony snap to the top of textual flagpoles. It's a remark laminated with intent, and it goes to the heart of Gibson and Sterling's alterations, for the talk, just before, has been of poetry. Their 1855 is indeed founded on an imbalance between reason and imagination; they've deliberately enthroned science at poetry's expense, starting with Byron's new career as politician and working on down through instance after instance in which those qualities that cannot be quantified or analysed have been scrubbed out of daily life. The result is polarization. Their London streets are so dark because they've become the repository of all unacknowledged desires. The repressed must return, and it does, it erupts in the form of a Blakean insurrection scattering prophetic gibberish through the poisonous fog. A world 'unnaturally' divided tries convulsively to reunite. Other imaginary histories register their own artificiality in passages of unease or in premonitory dreams sent by the author to warn a character that the world is less solid than they think. Quite frequently, the Caesar of a surviving Rome gets visions of a tumbled Colosseum: Lincoln, retired and disgraced in 1870, senses that he has outlived his proper span. The weirdest ripple of doubt probably passes across the Axis-occupied America of Philip K. Dick's *The Man in the High Castle*, when the yarrow stalks of the I Ching are made to deliver the plain message that the world is not true, an oracle very puzzling to the people of Dick's most tautly realized novel, who must continue to live all the same where the punctilious agents of the Greater East Asia Co-Prosperity Sphere rule San Francisco, delighting in Mickey Mouse watches and other remnants of Americana. It's as if the deviance of imaginary histories from the truth were a pervasive wrongness you could feel, infiltrating everything; as if what is not true was also out of true, subtly misshapen throughout its fabric.

No contradiction is necessarily involved when a genre devoted to change also throws up these momentary confessions of inadequacy. The characters may grow uneasy but the authors are not. Such nods back towards the unaltered world cheerfully acknowledge, in fact, a truth about the making of false realities. They are one manifestation of real history's paradoxical conservation by the process of chopping and changing it. Some authors like to conjure plots from a cod 'Law of the Conservation of History' which makes events right themselves no matter how hard you try to force a detour. That way they have their cake and eat it too; they get to tamper without disturbing a comfy Panglossian belief that everything is as it is for good reason. But a more fundamental kind of conservation is written deep into the structure of imaginary histories, and unites their ploys with the transactions that go on during the ordinary exercise of the historical imagination. Altered histories work with, work *on*, the sense of the past that readers bring to them. They operate upon what we think we know about Victorians or Romans or Christopher Columbus; their raw material is real history, not as real historians study it or understand it, but as a ready pack of characteristic images – the reader's mind's stock of associations between Victorians and crinolines, Romans and bloodstained togas, Columbus and cockleshell voyages off the map. They rely on the expectations they disrupt. History, joked *1066 and All That*, 'is what you can remember'. What you, the reader, can remember furnishes all that the imaginary historians have in the storecupboard once they have vanished the actual record. Of course they can invent; they can tell you things you didn't know; but they can never go beyond that point of invention where contact is lost with your original sense of history, or they also lose the point of making an alteration. If everything changes, then the changes are no longer measurable against the real past and can no longer be enjoyed. An imaginary history which forked off into a completely unrecognizable sequence of events would have no resonance at all.

And what first loaded meaning into the capsule props beloved of imaginary historians, except our customary understanding of history? Imaginary history needs the real past to instil such hybrids as the US Hitler stamp with reflexive, immediate significance. The same applies to the human population of altered worlds: the memory of real history acts as an essential preservative for the stated identities of the cast. The less 'Victorian' the things people are shown to be doing, for example, the more exaggeratedly Victorian in manners and outlook they have to be. Detached from their setting, they have to possess their Victorian identity as a concentrated essence of period, and the only source from which it can be pumped is our pool of shared expectations. So in undistinguished books you get an unusually high proportion of inevitable characterizations, characterizations done by numbers (though the numbers are scrambled). Predetermined bundles of mutton-chop whisker and check trousering masquerade as human. Even excellent authors in the genre cannot ever quite invent from scratch. Person building must always happen on a grid of references. The scope for surprise and sympathy that does remain stems from the diversity of references available, which can become a sort of palette. The inhabitants of invented worlds can be tinted to flesh colour with shades extracted from here, here and (unexpectedly) *here*. Gibson and Sterling were able to find out-of-the-way and almost novel stereotypes for *The Difference Engine*. They give us, in the underworld adventures of Mallory the savant, not the gestures and speaking voice of Victorian prudery (the popular choice of period trademark) but an active and half-shamed sexual hypocrisy like an out-take from *My Secret Life*. In the cloak-and-dagger part of the plot, they give us, rather than Sherlock Holmesery and predictable gaslit mayhem, an amalgam whose constituent parts could be listed like movie credits. Feel for detective procedure: Wilkie Collins. Criminal mastermind supplied by: Jules Verne. Action sequences: Spring-Heeled Jack of Limehouse. Apocalyptic lighting effects: the painter John Martin. Et cetera.

Still, the most obvious aspect of their 1855 is its premature, its blatantly precocious similarity to our own time. Their updating is not always smooth, and when Gibson and Sterling tweak over-hard to align 'clackers' with 'hackers', or to make London pavements into mean streets the *Neuromancer* audience will like, it can seem that the past is just being invested with arbitrary cool. Hey, these aren't dull Victorians. These are hip Victorians! Turbo Victorians! Victorians with Attitude! But modernizing the Victorians can also mean discovering a modernity in them that was really there. Gibson and Sterling are frivolous *and* serious here: travesty is mingled with an operation on our sense of the Victorian past that – like the slang 'Rads' – only looks like travesty. They have an agenda. Gibson, for one, has long been interested in pushing back the starting dates for the communications phenomena that fascinate him: he has argued before that access to the unsituated dimension of 'cyberspace' really began with the telephone, and now it looks as if the Morse key of a telegraph apparatus is being saluted as an even earlier gateway. He exaggerates, of course, but the effect of this kind of exaggeration is to render an element of the past visible. The same is true of *The Difference Engine* in sum; a wholesale device, after all, for solidifying theories, completing abandoned Engines, and turning wispy speculations into a visible manifold of events. More: it pursues a kind of secret, unlikely fidelity to elusive strains of genuine nineteenth-century technological feeling by exploiting our ability to recognize them as familiar once they have been transposed into current terms. The cartoonish enthusiasms for machinery in *The Difference Engine*, and the gothic apprehensions, are both blatant versions – therefore visible versions – of real feeling. The book is filled with disguised continuities. This is Gibson and Sterling's subtlest use of imaginary history's constant return upon the real. Imaginary history's prime material, the consensual vision of what used to be, is generally a sluggish thing, fond of its segmented sense of period, slow to register a new thought; but it can be hoodwinked into expansion. Gibson and Sterling's imaginary history returns the past to us, garishly refurbished, loudly wonderful; but those prove to be the terms on which, as when we look at the ranked cogs in South Kensington, we can glimpse the past's possibility alive.

If the Difference Engine re-enchants computers, *The Difference Engine* refreshes the nineteenth century. Let me give the best illustration. The Central Statistical Bureau of Gibson and Sterling's 1855 is a special effect, with its pyramid walls pierced by smokestacks, its polished halls of pandemonium within. The spinning Engines, multiplied in row upon row of 'majestic gearage' are 'like some carnival deception, meant to trick the eye'. In our 1855 a 'computer' was a person who did bulk arithmetic. But the unsettled admiration the Bureau provokes had an exact real-world counterpart. The *Quarterly Review* of June 1850 contained forty-six raptured pages devoted to the 'Mechanism of the Post Office'. There were no literal machines at all in the giant sorting halls the article explored, stacked one on top of another out of the reach of daylight – except a lift, described with estranged wonder as 'a very ingenious contrivance suggested by Mr Bokenham'. Instead they presented a 'mechanism' built from human components, each class of whom broke down the 'multitudinous mass' of the post by one specific stage: the red-coated postmen 'like a body of soldiers playing for their very lives at cards' who face all the 2,288,000 weekly letters in the same direction; the clerks who pass one verifying finger over each stamp with 'rapidity ... most astonishing'; the postmarkers whose right hands 'destroy from 6,000 to 7,000 Queen's heads in an hour'; the twenty-one sorting clerks through whose sets of fourteen tiny identical pigeonholes pass 'the whole of the correspondence of the United Kingdom, not only with itself, but with every region of the habitable globe, primarily arranged!' Except for the packages gingerly removed from circulation because they enclosed gunpowder or leeches or 'a human stomach &c', and the odd letter addressed only to 'sromfredevi' (Sir Humphry Davy), every item, rejoices the *Quarterly Review*, departs again in its right direction ·'by two great pulsations', 'diurnally

directed along six arterial railways to about 600 principal towns'. It
was all 'very strange' but 'magnificent'. Since the real years between
have abolished the strangeness, it takes the replacement of the GPO
by a mirage of Babbage Engines, the morphing of the human
circuitry into impossible metal, to remind us that what the *Quarterly
Review* is describing is – that thing whose promise comes and goes
in different lights – data processing.

Contributors

ISOBEL ARMSTRONG is professor of English at Birkbeck College, University of London. Her most recent book, *Victorian Poetry: Poetry, Poetics and Politics*, appeared in 1993. She is working on a book on glass and its meaning for the nineteenth century, which she thinks of as a kind of cultural poetics.

GILLIAN BEER is the King Edward VII Professor of Englsih Literature and President of Clare Hall, University of Cambridge. Among her books are *Darwin's Plots* (1983) and *Open Fields: Science in Cultural Encounter* (1996).

NEIL BELTON was born and educated in Dublin, first at a Catholic school which preferred Latin to biology and later at University College Dublin. He has been a commissioning editor and publisher in London since the early 1980s.

MARINA BENJAMIN is Arts Editor of *New Statesman and Society*. She has edited two collections of essays exploring women's relationship to natural knowledge and is currently writing a book about 'the end of the world'. (Producing this essay has forever changed her specular practices – she now needs glasses.)

PORTIA DADLEY has just completed a doctorate at York University on science and the nineteenth-century American imagination.

LAVINIA GREENLAW's most recent collection of poems, *Night Photographs*, was published by Faber in 1993. During 1995 she was Writer in Residence at the Science Museum, where she produced a sequence of poems in response to the Museum's collection.

ANNE JOSEPH is a graduate student pursuing a doctorate in Political Economy and Government at Harvard University and a law degree at Yale. She holds an M.Phil. in the History and Philosophy of Science from Cambridge University.

JON KATZ is a media critic. He has worked in newspapers and broadcasting and written six books – five novels and a non-fiction book, *Virtuous Reality: The War for America's Cultural Soul*, published by Random House in 1996. He has written for *Wired, Rolling Stone, New York* magazine, *Vogue* and *GQ*. His essay in this book was first printed in *Wired*.

ALEX SOOJUNG-KIM PANG received a Ph.D. in History and Sociology of Science from the University of Pennsylvania. He has been a postdoctoral student and lecturer at Stanford University, University of California–Berkeley and University of California–Davis.

TOM PAULIN's latest collection of poems, *Walking the Line*, was published by Faber in 1994. He teaches at Hertford College, Oxford, and is writing a critical study of William Hazlitt.

SIMON SCHAFFER is Reader in History and Philosophy of Science at the University of Cambridge. He has recently published on Babbage's calculating engines and the factory system and on Victorian standards of precision measurement.

FRANCIS SPUFFORD has edited *The Chatto Book of the Devil* and *The Chatto Book of Cabbages and Kings*, an anthology of lists used as a literary device. He reviews for *The Guardian*. Faber are publishing *I May Be Some Time*, a book about the British mythology of polar exploration, in 1996.

DORON SWADE, Senior Curator of Computing and Information Technology at the Science Museum, was an electronics design engineer before turning to the history of technology. He led the Museum's project to construct the Difference Engine No. 2, curated the Babbage bicentennial exhibition in 1991 and has written on Babbage and on the computer age (with Jon Palfreman).

JENNY UGLOW abandoned any attempt at a scientific education at thirteen, but became fascinated by Victorian science and technology while researching her books *George Eliot* and *Elizabeth Gaskell: A Habit of Stories*. She is now writing on William Hogarth.

ALISON WINTER did her undergraduate work at the University of Chicago, followed by an M.Phil. and Ph.D. at Cambridge in the history of science. Her doctoral dissertation focused on the cultural history of mesmerism in Victorian Britain, and is now being revised for publication by the University of Chicago. She is currently Assistant Professor in History at the California Institute of Technology.

Acknowledgements

We would all like to thank our editor, Julian Loose, for his support and interest. In addition, Marina Benjamin would like to thank the Scouloudi Foundation for their generous support of the research for her essay, and to thank Simon Schaffer, Alison Morrison-Low and Jenny Wetton for leads to the elusive Dancer. Anne Joseph and Alison Winter would like to thank Henry Atmore, Adrian Johns, Daniel Kevles, Mac Pigman, Simon Schaffer, James Secord and Andrew Warwick for helpful discussion of the issues, useful information and comments on early drafts of their paper. And Alex Soojung-Kim Pang thanks Robert Kohler, Henrika Kuklick, David Hollinger and Michael Dennis for their advice and encouragement, and Maurice Toler (North Carolina State University Archives), Helen Samuels (MIT Archives) and Karen Worley (Williams College Library) for their invaluable research assistance.

Index